Developments and Applications for ECG Signal Processing

Modeling, Segmentation, and Pattern Recognition

Developments and Applications for ECG Signal Processing

Modeling, Segmentation, and Pattern Recognition

Edited by

João Paulo do Vale Madeiro
Paulo César Cortez
José Maria da Silva Monteiro Filho
Angelo Roncalli Alencar Brayner

ACADEMIC PRESS
An imprint of Elsevier

Academic Press is an imprint of Elsevier
125 London Wall, London EC2Y 5AS, United Kingdom
525 B Street, Suite 1650, San Diego, CA 92101, United States
50 Hampshire Street, 5th Floor, Cambridge, MA 02139, United States
The Boulevard, Langford Lane, Kidlington, Oxford OX5 1GB, United Kingdom

Notices

Library of Congress Cataloging-in-Publication Data
A catalog record for this book is available from the Library of Congress

British Library Cataloguing-in-Publication Data
A catalogue record for this book is available from the British Library

ISBN: 978-0-12-814035-2

For information on all Academic Press publications
visit our website at https://www.elsevier.com/books-and-journals

Working together
to grow libraries in
developing countries

www.elsevier.com • www.bookaid.org

Publisher: Jonathan Simpson
Acquisition Editor: Glyn Jones
Editorial Project Manager: Peter Jardim/Joshua Mearns
Production Project Manager: Sruthi Satheesh
Designer: Victoria Pearson

Typeset by VTeX

Contents

João Paulo do Vale Madeiro, Paulo César Cortez,
José Maria da Silva Monteiro Filho and
Priscila Rocha Ferreira Rodrigues

João Paulo do Vale Madeiro,
José Maria da Silva Monteiro Filho and
Priscila Rocha Ferreira Rodrigues

João Paulo do Vale Madeiro,
José Maria da Silva Monteiro Filho and
Priscila Rocha Ferreira Rodrigues

Contributors

Angelo Roncalli Alencar Brayner

Department of Computing Science, Federal University of Ceara, Fortaleza, Ceará, Brazil

Paulo César Cortez

Department of Teleinformatics Engineering, Federal University of Ceara, Fortaleza, Ceará, Brazil

José Maria da Silva Monteiro Filho

Department of Computing Science, Federal University of Ceara, Fortaleza, Ceará, Brazil

João Paulo do Vale Madeiro

Institute for Engineering and Sustainable Development (IEDS), University for the International Integration of the Afro-Brazilian Lusophony – UNILAB, Redenção, Ceará, Brazil

Olavo Luppi Silva

Center of Engineering, Modeling and Applied Social Sciences, Federal University of ABC, São Paulo, Brazil

Roberto Coury Pedrosa

University Hospital Clementino Fraga Filho, Federal University of Rio de Janeiro, Rio de Janeiro, Rio de Janeiro, Brazil

Priscila Rocha Ferreira Rodrigues

Department of Computing Science, Federal University of Ceara, Fortaleza, Ceará, Brazil

João Loures Salinet Jr.

Center of Engineering, Modeling, and Applied Social Sciences, Federal University of ABC, São Paulo, Brazil

Preface

The electrocardiogram (ECG) is considered the world's reference standard for the noninvasive diagnosis of cardiac arrhythmias and conduction disorders. The digital processing of this signal, aiming at the automatic extraction of parameters and the recognition of patterns of cardiac diseases, has become widely popular as a tool to aid medical diagnosis in cardiology. The automatic extraction of ECG features comprises a set of steps, starting from the detection and segmentation of its characteristic waves and the determination of the series of corresponding intervals, and proceeding to the processing of time series derived from the ECG, such as heart rate variability. The wide range of factors that can be extracted from the ECG signal processing includes the metrics classically employed in arrhythmia detection, such as characteristic wave lengths (PQRST complex), amplitudes, and intervals or segments between the (beginning/end) boundaries of waveforms, signal content analysis in both time/frequency (wavelet transform) domains, and metrics derived from the mathematical modeling of waveform morphologies, through parameterizations of known mathematical functions.

Our goal in writing this book is to enhance the importance of accurate and reliable techniques for ECG signal processing, which has the potential to significantly increase the applicability of ECG use for diagnostic purposes, and expose, and detail a wide range of challenges in the processes of acquisition, preprocessing, segmentation, mathematical modeling, and pattern recognition in ECG signals, while presenting, in each case, practical and robust solutions based on digital signal processing techniques. Computational algorithms, based on digital signal processing techniques, are essential for the automatic analysis of ECG signals in the most diverse scenarios: clinical investigation, telemonitoring, postoperative follow-up, monitoring of patients in intensive care, heart rate variability analysis, prediction of adverse events, among other uses. Parameters extracted through signal segmentation and analysis serve as a basis for statistical or artificial intelligence-based classifiers to recognize patterns of heart diseases or other adverse events. Despite the fact that many approaches and methods have already been published in the literature, mainly concerning QRS complex detection, and also (but at a lower scale) P- and T-wave delineations, as well as the extraction of ECG signal parameters, such methodologies focus on clean clinical signals, and have their validation based on a restricted group of characteristic waveform morphologies.

Therefore dealing with ECG processing and feature extraction, and considering a signal, which is recorded from the patient being immersed in his/her daily routine (as it is the case of the tele-health devices) or even affected by certain heart diseases that cause changes or even suppressions of the characteristic waves, the need for new and robust research in this area is evident.

Thus this book has the potential to contribute to research regarding automatic analysis of ECG signals, extending the resources for rapid and accurate diagnoses, especially in long-term signals, in which the visual analysis by specialist physicians becomes unviable, in addition to motivating the development of alert systems/equipment for patients with critical heart disease.

The editor João Paulo do Vale Madeiro wishes to thank CNPq, the Brazilian Research Council, for the source of funding via Grant No. 426002/2016-4.

João Paulo do Vale Madeiro
Paulo César Cortez
José Maria da Silva Monteiro Filho
Angelo Roncalli Alencar Brayner

Editors
May 2018

Classical and Modern Features for Interpretation of ECG Signal

**João Paulo do Vale Madeiro*, Paulo César Cortez†, João Loures Salinet Jr.‡,
Roberto Coury Pedrosa§, José Maria da Silva Monteiro Filho¶,
Angelo Roncalli Alencar Brayner¶**

**Institute for Engineering and Sustainable Development (IEDS), University for the International Integration of the Afro-Brazilian Lusophony – UNILAB, Redenção, Ceará, Brazil
†Department of Teleinformatics Engineering, Federal University of Ceara, Fortaleza, Ceará, Brazil
‡Center of Engineering, Modeling, and Applied Social Sciences, Federal University of ABC, São Paulo, Brazil
§University Hospital Clementino Fraga Filho, Federal University of Rio de Janeiro, Rio de Janeiro, Rio de Janeiro, Brazil
¶Department of Computing Science, Federal University of Ceara, Fortaleza, Ceará, Brazil*

1.1 ELECTRICAL ACTIVITY OF THE HEART

The heart, or the cardiac muscle, has the function of pumping blood to the organs and specifically to the lungs, facilitating the exchange of gases, absorbing oxygen and eliminating carbon dioxide. It is divided into upper (atria) and lower (ventricles) chambers, and it may also be considered to be composed of two separate pumping systems: a right heart that pumps blood through the lungs, and a left heart that pumps blood through the peripheral organs. Each atrium helps to move blood into the ventricle, and the ventricles supply the main pump force needed to push blood through the pulmonary and peripheral circulation systems (Hall, 2011), see Fig. 1.1 (this figure was published in Textbook of Medical Physiology, Arthur C. Guyton and John E. Hall, Chapter 9: Heart Muscle; The Heart as a Pump and Function of the Heart Valves, Page 104, Copyright Elsevier Inc. (2006)).

In addition to the atrial and ventricle muscles, myocardial tissue consists of excitable and contractile fibers capable of developing self-regenerative electrical activity, presenting specific functions of generation, conduction, and contraction, depending on its anatomical location. There are two basic groups of cells in the myocardium that are important for cardiac function.

Developments and Applications for ECG Signal Processing. https://doi.org/10.1016/B978-0-12-814035-2.00007-4

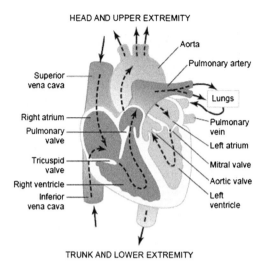

FIGURE 1.1

Structure of the blood pumping system of a human heart concerning heart chambers and heart valves.

1.1.1 CELLS RESPONSIBLE FOR MYOCARDIAL CONTRACTION

The contractile apparatus of cardiac fibers consists of a complex of contractile proteins, composed of actin, myosin, tropomyosin, and troponin, which—in the presence of calcium and adenosine triphosphate—interact with each other, causing contraction. Such cells possess the property of contractility, that is, the ability to shorten and then return to their original length. For a myocardial cell to contract, the cell membrane must be discharged electrically (a process called depolarization), causing a change in electrical charge across the membrane, resulting in the flow or movement of certain ions (especially sodium). The depolarization process also allows the entry of calcium into the cell, where it is responsible for the binding part between actin and myosin of the sarcomere (basic contractile unit of myocardial fibers), resulting in contraction.

1.1.2 THE ELECTRICAL SYSTEM

The cells that make up the electrical system of the heart are responsible for the formation of the electric current and the conduction of this impulse to the contractile cells of the myocardium, where the depolarization activates the contraction. Certain cells in the electrical system have the ability to generate an electrical impulse (a property referred to as spontaneous automaticity or depolarization). Cells possessing this property are known as "pacemaker cells". These cells are found in the sinus node, in the cells responsible for atrial conduction, in the area immediately above the atrioventricular (AV) node, in the low portion of the AV node, in the His bundle, and in the Purkinje ventricular system.

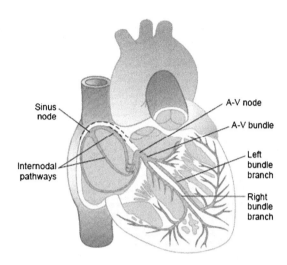

FIGURE 1.2

Schema of the specialized excitatory and conductive system of the heart that controls cardiac contractions.

As illustrated in Fig. 1.2 (this figure was published in Textbook of Medical Physiology, Arthur C. Guyton and John E. Hall, Chapter 10: Rhythmical Excitation of the Heart, Page 116, Copyright Elsevier Inc. (2006)) the sinus node or sinoatrial (S-A) node, in which the normal rhythmical impulse is generated, has the highest rate of spontaneous depolarization and acts as the primary pacemaker. Internodal pathways conduct the impulse from the sinus node to the AV node. The AV bundle conducts the impulse from the atria into the ventricles, and the left and right bundle branches of Purkinje fiber conduct the cardiac impulse to all parts of the ventricles (Wasilewski and Poloński, 2012).

1.1.3 BASIC ELECTROPHYSIOLOGY

To better comprehend the disturbances in the electrical activity of the heart, it is important to have a basic knowledge of the electrical properties of the contractile myocardial cells and of the pacemaker cells (Fig. 1.3). With the use of a single microelectrode, the electrical event (or action potential) of a single cell can be recorded.

The electrical activity of the cardiac fiber depends initially on the intracellular and extracellular concentration of alkaline, alkaline earth, and halogen ions. In the cellular protoplasm, low concentrations of sodium ions and high concentrations of potassium and chloride ions are maintained. The calcium ion undergoes variations in its intracellular concentration, depending on the muscular concentration. These same ions are present in the liquid that surrounds the cells, and the sodium and calcium concentrations are higher, the potassium lower, and the chlorine isotonic with the plasma. In the absence of contraction, the distribution in different concentrations

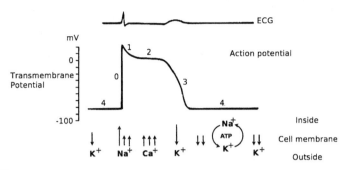

FIGURE 1.3

Schematic representation of the action potential of a ventricular myocardial cell, where the arrows indicate the movement of ions through the membrane.

intracellular and extracellular of these ions, separated by the cellular membrane, induces the formation of transmembrane ionic electrical potential known as Nernst potential, characteristic of each ionic species. The quantitative sum of the potentials of the involved ionic species generates a transmembrane potential difference that, at rest, is negative on the internal side in relation to the external face.

The transfer of sodium, potassium, and calcium ions from the medium into the extracellular medium, and vice versa, depends on the selective permeability of the cell membrane to each of these elements, which depends on the action of proteins specialized in the passage of these ions through the cell membrane.

To perform its function, the heart must undergo cyclical variations in transmembrane potential. These variations define the cardiac cycle or the set of electrical and mechanical activities of the heart, comprised between two successive contractions.

During the phase of the cardiac cycle in which the fibers are at rest, the distribution of those ions generates a stable transmembrane spontaneous potential difference called the resting potential, which is established around −90 mV under normal conditions (Fig. 1.3). In other words, the inside of the cell membrane is electrically negative compared to the exterior of the cell membrane. This is due to a distribution of ions across the cell membrane. During this phase (phase 4), the cell membrane presents a peculiar characteristic, in which the conductance for the potassium ion is about 10 times greater than that for other ions, allowing the equilibrium potential of this ion to be the potential of rest. At this stage, Na+ is found at high concentrations outside the cells and at low concentrations inside the cells. Because of this Na+ concentration gradient, there is a physical force generated in order to cause Na+ to enter the cell. However, to maintain this chemical gradient, there is an energy expenditure, that is, there is an opposite electric gradient. Moreover, during this initial phase of action potential, the cell membrane is impermeable to Na+. The K+ is found at high concentrations inside the cell and at low levels outside the cell, and the mostly open channel is the potassium channel. This ion, in small quantity, is able to cross the membrane.

During phase 4, therefore, K+ is able to traverse the cell membrane from the inside out. Due to this direction of flow or movement of K+, the inside of the cell becomes electrically negative in relation to the outside, which becomes positive. Therefore the resting potential is primarily dependent on the concentration gradient of the K+ ion across the cell membrane.

To initiate depolarization (Fig. 1.3), a complex mechanism of ion channels (rapid channel) in the cell membrane opens momentarily (duration is approximately 1 ms), allowing a rapid Na+ entry into the cell through its gradient of concentration. Since there is now a flow of positively charged ions into the cell, it becomes electrically positive (approximately +20 mV), while the outside of the cell becomes negative. This part of the action potential is called phase 0 (zero).

The ability of cardiac muscle cells to drive the electrical impulse that triggers contraction, with adequate strength and speed, depends fundamentally on the resting potential. Hence the importance of maintaining the resting potential within a narrow range of variation. The depolarization or decrease (in absolute value) of the resting potential is frequently associated with conduction and rhythm disturbances, as well as the reduction of cardiac pumping efficiency.

When phase 0 occurs in the ventricular muscle cell, at the same time, the QRS complex on the electrocardiogram is recorded. The P wave is generated in phase 0, when it occurs in the atrium muscle cell. As the ion channel mechanisms gradually close, and the Na+ input progressively decreases its concentration gradient, the internal electric charge becomes less positive, thus initiating the repolarization process (phase 1). During phase 2, the action potential is approximately isoelectric, and the cell membrane remains depolarized. At this point, a small amount of Na+ is entering through the fast channel, while Ca++ and, possibly, a smaller amount of Na+, are entering through the slow channel. Phase 2 of the ventricular myocardial cell occurs at the time that the ST segment is being recorded on the electrocardiogram.

Phase 3 represents rapid repolarization, during which the interior of the cell becomes negative again. This is caused by increasing the output or movement of the K+ from the inside to the outside of the cell. Stage 3 in the ventricular myocardium cell occurs during T-wave recording on the electrocardiogram (Fig. 1.3). Repolarization is complete at the end of phase 3. The interior of the cell is again approximately −90 mV.

However, the distribution of ions across the cell membrane is different from that immediately prior to the onset of depolarization. Due to the entry of Na+ into the cell and the loss of K+ from the cell, there is a high concentration of intracellular Na+ and a low concentration of intracellular K+. This could cause the cell to depolarize a second time, but repeated depolarizations without proper redistribution of Na+ and K+ would eventually lead to worsening of cellular function. Then, during phase 4, a special pump mechanism in the cell membrane is activated. This transports Na+ from the inside to the outside and brings the K+ back into the cell. This mechanism is dependent on ATP (Adenosine Triphosphate) as an energy source.

The level of the resting membrane potential (phase 4) for depolarization initiation is an important determinant of conductivity (the ability to cause a cellular uptake to depolarize, and the rate at which the clustered cells are depolarized) of an electrical impulse from one cell to another. For a minus negative resting potential at the start

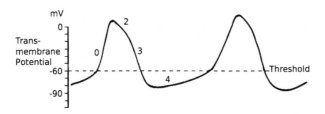

FIGURE 1.4

Schematic representation of an action potential of a pacemaker cell (sinus node).

of phase 0 (for example, −60 mV as opposed to −90 mV), the lower the phase-shift increase rate of the action potential. Conductivity is directly related to the rate of increase of phase 0 of the action potential. The factors that determine the rate of increase of phase 0 (i.e. conductivity) are the gradient of Na+ concentration at the beginning of phase and the gradient of K+ during phase 4. For example, an increase in extracellular K+ will result in a decrease in the K+ gradient, resulting in a decreased resting potential.

The heart should contract around every 0.75 seconds at rest and up to three or more times per second during very intense workouts. But who guarantees this rhythm? How is the regulation made to adapt its functioning to the metabolic needs of the organism?

The heart rhythm is generated by the heart itself, so that an isolated heart can be kept functioning for hours in vitro if we provide an ionic medium with sufficient nutrients and oxygen for its metabolism. In the heart of a mammal, this rhythm is generated by modified muscle cells from a region in the right atrium called the si-nus node. Cells in this region have a particularity: they do not have a stable resting potential like the cells of the ventricular musculature. After a potential action, the repolarization takes the transmembrane potential to approximately −60 mV (com-pared to −90 mV of the ventricular cells). Unlike the ventricular cell, in which the potential remains constant in the diastole (during which the cardiac muscle is relaxed and ventricular filling occurs), a slow depolarization occurs in the sinus node cells, which, after a certain time, triggers a new potential for action. They therefore are able to self-stimulate, thanks to the slow depolarization that occurs in the diastole. We say that the sinus node is the heart pacemaker. The action potential of a "pacemaker" cell differs significantly from the potential of a contractile myocardial cell (Fig. 1.4).

Once the action potential has been generated in the sinus node, this potential is propagated to neighboring cells, invading atrial muscle tissue and activating both atria (Fig. 1.2). On the way, it also activates the atrioventricular conduction system, also formed by modified muscle cells, and, through it, the whole ventricular muscular mass. This is done in such a way that each time an action potential is generated in the sinus node, as a consequence, ventricular activation occurs, making the heart to function as a whole.

The propagation of the action potential from one cell to another is ensured by the existence of specialized structures of the plasma membrane, the communicating

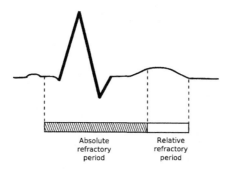

Absolute
refractory
period

Relative
refractory
period

FIGURE 1.5

Scheme of the electrocardiogram refractory period.

junctions, which allow the establishment of an ionic current flow between contiguous cells, providing an electrical continuity between them.

The curve in phase 4 of the action potential of a pacemaker cell (Fig. 1.4) is an important factor in impulse formation. For a larger slope, the faster the frequency of pulse formation, for a smaller slope, the slower the frequency of pulse formation. Activation of the sympathetic nervous system makes the inclination greater, thus increasing automaticity. Stimulation of the parasympathetic nervous system will have an opposite effect. Commonly used antiarrhythmic drugs may decrease the rate of spontaneous depolarization (although this is not the only mechanism of action of antiarrhythmic drugs).

Clinically, the most important group of pacemaker cells are those found in the sinus node, AV node, and ventricular conduction system. The trigger frequency (spontaneous depolarization) differs at each of these sites. The sinus node is the heart's primary pacemaker and has a normal firing frequency between 60–100 bat/min.

The firing rate of the atrioventricular node is 40–60 bat/min, and for the ventricular conduction system (His-Purkinje system), it is less than 40 bat/min. This decremental decrease has important physiological implications. The lower the pacemaker cell location, the harder it is to reach the potential threshold, that is, the pacemaker cells are deprived of spontaneous depolarization. Thus the "pacemaker cells" in the AV and ventricular nodules are "escape pacemakers"; they do not spontaneously produce an electrical impulse unless the faster pacemaker (e.g., sinus node) fails. Therefore if the sinus rate drops significantly below 60 bat/min, the AV node escape should resume. Similarly, if a supra-ventricular impulse does not reach the ventricle within approximately 1.5 s (equivalent to a frequency of 40 bat/min), the ventricular system, i.e., ventricular escape, should resume. However, the rate of this "pacemaker-escape" rhythm may increase or decrease in several diseases, with drugs, sympathetic or parasympathetic stimulation.

Another important concept is the refractory period (Fig. 1.5). The refractory period of the ventricle begins with the onset of phase 0 (the onset of the QRS complex) and ends at the end of phase 3 (or T wave). This can be more conventionally divided into two parts: (1) absolute refractory period, which starts at phase 0 and ends at the middle of phase 3 (approximately at the apex of the T-wave); (2) relative refractory period, which extends through the remainder of phase 3 (until the end of the T-wave). During the absolute refractory period, the cell does not propagate or conduct the action potential. During the relative refractory period, a strong stimulus may result in a stimulus propagation but not necessarily normal action potential.

1.1.4 PULSE FORMATION MECHANISMS
There are two basic mechanisms related to the origination of an electrical impulse in the myocardium:

1) Automaticity: the impulse can be originated through the mechanism of automaticity already described; 2) Re-entry: the basic components of the re-entry mechanisms include two conduction pathways, one of which has unidirectional locking (or a long refractory period), and the other has a slow conduction, where the transmission time around the circuit is large enough for the refractory period of the Purkinje cells to recover, thereby making a circular transmission circuit.

1.1.5 CONDUCTION OF THE CARDIAC IMPULSE
As already explained, the normal cardiac impulse originates from the sinus node. The conduction from the sinus node occurs through three internodal paths. The velocity of conduction through the atrium is approximately 1000 mm/s. The atrioventricular node is located inferiorly in the right atrium, anterior to the ostium of the coronary sinus, and above the tricuspid valve (Fig. 1.2). The rate of conduction through the AV node is slow (approximately 200 mm/s). The AV node is a complex fiber network. These fibers converge to the lower margin forming a discrete beam of fibers, the bundle of His. This structure penetrates the fibrous ring and reaches the superior margin at the interventricular septum in its muscular portion, where it gives rise to the branches. The left branch has a complex and variable anatomy. However, two groups may be considered: superior, previously located, and inferior, posteriorly located to the interventricular septum. The Purkinje network of the interventricular septum can be originated as a separate radiation or as fibers, either from the anterior or posterior beam radiation of His.

The right branch runs down the right side of the interventricular septum. As soon as the electric impulse reaches the AV node, it passes into the His bundle and then descends to the Purkinje-beam branches simultaneously. The first part of the ventricle to initiate depolarization is the middle portion of the left ventricular septum. The free wall of both ventricles are depolarized simultaneously. The speed of driving through the Purkinje network is fast, approximately 4000 mm/s.

1.1.6 HEART RATE CONTROL
The frequency with which the heart is fundamentally dependent is related to the triggering frequency of the action potential for the sinus node and its ability to spread

throughout the heart. However, as seen on a daily basis, a variable heart rate adapts to various conditions. This regulation is made by the influence of the hormonal and nervous systems on the activity of the sinus node pacemaker. There is, in fact, a balance between the activating action of the sympathetic and parasympathetic nervous systems on the sinus node.

During a race, for example, as the demand for oxygen and energy sources increases in the muscles, there is an increase in blood flow, as a result of the acceleration of heart rate and the force of myocardial contraction. At rest, a decrease in blood flow occurs because of decreased heart rate and contraction force. These changes are produced by the action of hormones, such as adrenaline released by the adrenal gland and carried to the heart by the circulation and neurotransmitters, and noradrenaline released by nerve endings of the sympathetic nervous system that innervates the heart. Adrenaline increases heart rate by accelerating the slow diastolic depolarization of the sinus node. This action is done by increasing the density of current pacemaker and calcium by adrenaline, through the phosphorylation of these channels.

Bradycardia (decreased heart rate) is produced by the reduction of sympathetic activity, associated with exacerbation of parasympathetic activity. The vagus nerve (the parasympathetic pathway for the heart) releases another type of neurotransmitter, acetylcholine, at its terminals. This substance retards diastolic depolarization through its depressant action on the current pacemaker, and also by activating a potassium channel through its interaction with muscarinic receptors, which are membrane proteins located in the vicinity of that channel. As a consequence, the cardiac pacemaker's firing frequency decreases.

Thus the perfect functioning of the heart depends on a coordinated sequence of events that occur in the plasma membrane, that is, depolarization in the sinus pacemaker, its propagation to the atria, to the atrioventricular conduction system, and to the ventricular musculature. In each of these stages, different types of ionic channels present in the cell membrane undergo changes that regulate the flow of different ions in a coordinated way, provoking the adequate electrical activation of each one of the regions of the heart, fundamental for the coordinated contraction of the heart.

1.2 THE ELECTROCARDIOGRAM SIGNAL

The ECG signal is a record of the potential differences produced by the electrical activity of the heart cells. The body by itself acts as a giant conductor of electrical current and any two points in the body can be connected by electrical electrodes to record an ECG or monitor the heart's rhythm.

The trace measured and recorded using suitable equipment refers to the electrical activity of the heart and forms a series of waves and complexes that have been arbitrarily called P wave, QRS complex, T wave, and U wave. The waves or deflections are separated by regular intervals (see Fig. 1.6).

The depolarization of the atrium produces the P wave; depolarization of the ventricles produces the QRS complex. Ventricular repolarization causes the T wave. The

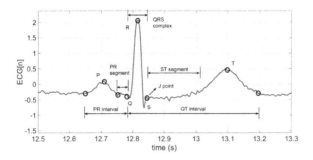

FIGURE 1.6

Characteristic waves and some basic parameters for ECG.

significance of the U wave is uncertain, but may be due to repolarization of the Purkinje system (Hall, 2011). The PR (iPR) interval extends from the beginning of the P wave to the beginning of the QRS complex. This should not exceed 0.20 seconds measured on ECG paper, where each small square represents 0.04 seconds. The upper limit of the normal duration of the QRS complex is < 0.12 seconds. A duration less than 0.12 seconds means that the impulse was initiated at or above the AV node (supraventricular). A duration of QRS > 0.12 seconds may mean an impulse originating from the ventricle or originating from the supraventricular tissue but with prolonged conduction through the ventricle. The T-wave is the ventricular recovery or repolarization of the ventricles.

The key to the interpretation of cardiac arrhythmia is the analysis of the interrelationship between the P wave, the PR interval (iPR), and the QRS complex, including its duration and configuration. The ECG should be analyzed with respect to the frequency, rhythm, locus of the dominant pacemaker, and the configuration of the P wave and the QRS complex. The relationship between ECG and cardiac anatomy is shown in Fig. 1.7.

The middle line of the diagram presented in Fig. 1.7 is the His bundle, dividing into the two branches of the conduction system. Any dysfunction above this point affects the P wave and the PR interval, whereas below this level it will affect the QRS complex. Some important events, which can be observed from the morphology or time duration of P-wave, PR interval, and QRS complex are highlighted below:

- P-wave: If for some reason the sinus node fails as a normal pacemaker cell, another atrial focus may take over and then the P-wave may have changes in its morphology. Alternatively, a second focus may be activated, that is, a secondary focus of the pacemaker (for example, the AV node) such that we will have an escape rhythm;
- PR-interval: when conduction through the atrium, AV node or His bundle is slow, the PR interval is increased. Changes in AV node conduction are the most common causes of increased iPR;
- QRS complex: if there is a delay or interruption in conduction within the branches, the QRS complex will typically widen. That is, the right bundle branch block or

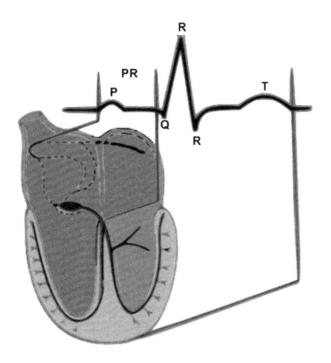

FIGURE 1.7

Relationship of the electrocardiogram to the anatomy of the conduction system.

left bundle branch block. An ectopic focus that initiates a pulse in the ventricle may also alter the shape of the QRS complex. When the ectopic focus originates above the bundle of His or not, but not in the branches, the ventricles are activated normally, and the QRS complex will remain the same, assuming that there is no delay in conduction in the branches. If the focus originates below, the QRS complex will widen due to different driving sequences.

Nowadays, there are many approaches to record ECG signal. According to the taxonomy of the state-of-the-art of the ECG measurement methods provided by da Silva et al. (2015), the several approaches can be divided into three classes: in-the-person, on-the-person, and off-the-person. Concerning the in-the-person category, there are equipments designed to be used inside human body, such as surgically implanted ones, subdermal applications or even ingested in the form of pills. These devices are used when less invasive approaches are not applicable or effective (Luz et al., 2016).

Devices in the off-the-person category are designed to measure ECG without skin contact or with minimal skin contact. According to da Silva et al. (2015), unlike the on-the-person methods, in which the user needs to wear the sensor, having it in contact with his body, in this case the sensor is integrated in a pervasive manner. As an example, the authors cited contactless systems based on capacitive sensors, which measure the small time-varying electric fields associated with the bioelectric activity of the heart. These sensors do not require direct contact with the body of the user, and can be designed to measure ECG at distances of 1 cm or more, even with clothing between the body and the sensor.

Contrasting with the off-the-person category, devices in the in-the-person category are placed inside the body of the person, and are generally used in extreme clinical scenarios to monitor or address medical conditions. One class are the implantable systems, of which artificial cardiac pacemakers are the most widely known example. Finally, the on-the-person category constitutes the most common approaches to ECG measurement used nowadays, which work by attaching a device, or some of its components, externally to the body surface. As classes of examples, we can cite bedside monitors for medical use and ambulatory systems, which include Holter monitors and, more recently, smart t-shirt and other wearable form factors.

In the next chapter, we will present a detailed description of the state-of-the-art ECG signal acquisition techniques, as well as present new developments, considering the collection of the raw signal by biomedical sensors to accessing digitized signal in computational environment for visualization and application of preprocessing algorithms.

1.2.1 DIPOLO FORMATION

According to Gari et al. (2006), the simplest mathematical model for relating the cardiac electrical activity to the body surface potentials is the single dipole model. This model has two components, a representation of the electrical activity of the heart, and the geometry and electrical properties of the body. Concerning the representation of the electrical activity of the heart, as an action potential propagating through a cell, there is an associated intracellular current and, consequently, a propagation in the direction of the interface of resting and depolarizing tissue, which is the elementary electrical source of the surface ECG, referred to as the current dipole. Considering that there is also an equal extracellular current flowing against the direction of propagation, charge is conserved and all current loops in the conductive media close upon themselves, forming a dipolo field. Actually, bioelectricity normally flows in current loops, as illustrated in Fig. 1.8. A loop includes four segments: an outward traversal of the cell membrane, an extracellular segment, an inward traversal of the cell membrane, and an intracellular segment. Each of the segments has aspects that give it a unique importance to the current loop.

Therefore the heart's total electrical activity at any instant of time may be represented by a distribution of active current dipoles. For modeling purposes, we can consider that all the individual current dipoles originate at a single point in space, and the total electrical activity of the heart may be represented as a single equivalent dipole, whose magnitude and direction is the vector summation of all the individual dipoles. Naming this resultant dipole moment as the heart vector $M(t)$, as each wave of depolarization spreads through the heart, the resultant vector changes in magnitude and direction as a function of time.

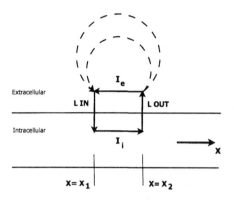

FIGURE 1.8

Flow of bioelectricity in current loops presenting four segments: an outward traversal of the cell membrane, an extracellular segment, an inward traversal of the cell membrane, and an intracellular segment.

The configuration of surface distribution of currents and potentials depends on the electrical properties of the torso. As a reasonable approximation, the dipole model can consider the body as linear, isotropic, homogeneous, with conductivity σ, and spherical conductor of radius R. Considering that the source for surface potential is represented as a slowly time-varying single current dipole located at the center of the sphere, Laplace's equation may be solved to give the potential distribution on the torso as (Gari et al., 2006)

$$\phi(t) = \cos\theta(t) \cdot 3|M(t)|/4\pi\sigma R^2. \tag{1.1}$$

The angle $\theta(t)$ is the angle between the direction of the heart vector $M(t)$ and the lead vector joining the center of the sphere to the point of observation or measure. Therefore the potential difference between two points on the surface of the torso is given as (Gari et al., 2006)

$$V_{AB}(t) = \mathbf{M}(t) \cdot \mathbf{L}_{AB}(t), \tag{1.2}$$

where $L_{AB}(t)$ refers to the lead vector connecting different points A and B of observation on the torso.

1.2.2 ECG LEADS

Electrocardiogram is traditionally recorded from two electrodes located on different sides of the heart. As a result of electrical activity of the myocardium cells, current flows within the body and potential differences are established on the body surface. One simple physical and mathematical model relating generation and propagation of the action potentials to the body surface potential differences is the dipole model, as explained in the previous section. The resultant electrical activity of the heart may be represented as a time-dependent heart vector, changing in magnitude and direction as the wave of depolarization spreads through the heart.

Therefore each ordered event related to the cardiac cycle provides a specific behavior for the resultant vector. Firstly, when atria depolarize, wave of depolarization descends through both atria. The vector can be decomposed into components pointing down, the most predominant component to the left and slightly to the back. When electrical impulse passes through the AV node, there is no measurable electrical activity at the body surface, which is referred to delay at the AV node. Subsequently, electrical activity depolarizes the His bundle and bundle branches. Then, the septal depolarization. As the action potential wave enters the septal myocardium, it tends to propagate left to right. Therefore the resultant heart vector points to the right. Next, apical depolarization, when the resultant vector points toward the apex of the heart which predominantly points down, to the subject's left, and slightly anterior. Subsequently, we observe left depolarization, for which we observe electrical activity for both left and right ventricle, but predominantly in the much more massive left ventricle. After the whole myocardium depolarizes, it starts to contract, and there is a period when no action potential propagates (absolute refractory period), and hence there is no measurable cardiac vector. After that plateau period, the cells begin to repolarize and another wave of charging spreads through the ventricles, which makes the heart return to its resting state (ventricular repolarization). As both the polarity and the direction of propagation of the repolarizing phase are reversed from those of depolarization, T-waves (related to repolatization waves) on the ECG are commonly of the same polarity as QRS complexes (related to despolarization waves) (Hall, 2011).

The time-dependent heart vector, with variable direction and magnitude, is projected onto 12 different lines with well-defined orientation, which are called 12 leads. Each lead provides a measure for the magnitude of the cardiac vector in a specific direction at each instant of time. The first important set of leads are the so-called standard bipolar limb leads, which comprises lead I, lead II, and lead III. For each lead, the electrocardiogram is recorded from two electrodes located on different sides of the heart, example, on the limbs. Electrodes are placed on each of the four extremities of a person. Each pair of electrodes make a closed circuit between the body and the electrocardiograph. Each bipolar limb lead is configured as follows:

- Lead I: The negative terminal of the electrocardiograph is connected to the right arm and the positive terminal to the left arm;
- Lead II: The negative terminal is connected to the right arm and the positive terminal to the left leg;
- Lead III: The negative terminal is connected to the left arm and the positive terminal to the left leg.

According to Einthoven's triangle model, drawn around the area of the heart, the two arms and the left leg form apices of a triangle surrounding the heart. The two vertices at the upper part of the triangle represent the connecting points between the two arms and the fluids around the heart, and the lower vertex is the point at which

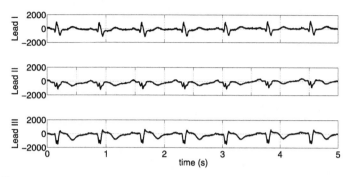

FIGURE 1.9

Electrocardiogram recorded from the three standard electrocardiographic limb leads: patient diagnosed with myocardium infarct.

the left leg connects with the fluids. Based on the Einthoven's model, if the electrical potentials of any two of the three limb leads are known at any given instant, the third one can be determined mathematically by the vectorial summation (Hall, 2011).

As an example, Fig. 1.9 illustrates an excerpt of the three bipolar limb leads from an electrocardiogram public available on PTB Database: patient001/s0010re (Goldberger et al., 2000). The corresponding person, a woman, 81 years old, had an infero-latera myocardial infarction. We can observe, with careful measurements and observance of polarities, that the signal for lead II can be obtained by the simple summation between the signal for lead I and the signal for lead III. Furthermore, the signal for lead I can be obtained by the summation between the signal opposed version for lead III and the signal for lead II, and the signal for lead III can be obtained by the summation between the signal for lead II and signal opposed version for lead I.

The precordial leads or chest leads report heart activity on the horizontal plane, therefore perpendicular to the plane focused by limb leads. This requires that six electrodes are placed around the torso, example, on the anterior surface of the chest, directly over the heart, at one of the points shown in Fig. 1.10.

Each electrode placed at each one of the referred points are connected to a specific positive terminal of the electrocardiograph, and the negative electrode, called central terminal. The same principle applies for the set of chest leads, connected on the basis of equal electrical resistances to the right arm, left arm, and left leg all at the same time. The different recordings for the precordial leads are known as leads V1, V2, V3, V4, V5, and V6.

Finally, for the augmented unipolar limb leads, two of the limbs (extremities) are connected on the basis of electrical resistances to the negative terminal of the electrocardiograph, and the third limb is connected to the positive terminal. We can interpret this configuration as the measure of a potential at a given extremity with respect to the average of the potentials of the other two extremities.

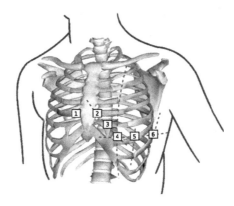

FIGURE 1.10

Scheme for placing electrodes aiming at recording precordial leads: V1, V2, V3, V4, V5, and V6.

Each augmented bipolar limb lead is configured as follows:

- Lead aVR: Positive terminal is on the right arm, and the electrodes placed on left leg and left arm are connected to the negative terminal (average of potentials at left leg and left arm);
- Lead aVL: Positive terminal is on the left arm, and the electrodes placed on left leg and right arm are connected to the negative terminal (average of potentials at left leg and right arm);
- Lead aVF: Positive terminal is on the left leg, and the electrodes placed on left arm and right arm are connected to the negative terminal (average of the arm leads).

1.3 ECG INTERPRETATION ON HEART RHYTHMS AND CONDUCTION DISORDERS

1.3.1 BRADYARRHYTHMIAS

Sinus Bradycardia

It is characterized by a heart rate slowing at 60 bpm or lower due a reduction of SA node firing automaticity or abnormal vagal tone (Pastore et al., 2009; Lilly, 2012). The characteristics of the ECG undergoing this arrhythmia are a P-wave positive at leads I, II, and avF, with heart rate between 50 bpm to 60 bpm (Pastore et al., 2009). Sinus bradycardia is normally asymptomatic and presented mostly in young population and athletes, but protuberant heart rate reduction can cause fatigue and syncope (Pastore et al., 2009; Lilly, 2012).

An excerpt of ECG signal showing an episode of Sinus Bradycardia is presented in Fig. 1.11. This episode refers to the record 232 from MIT–BIH Arrhythmia Database (Goldberger et al., 2000).

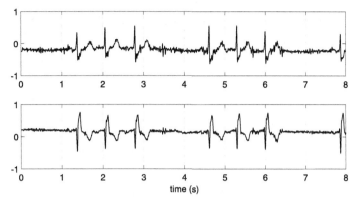

FIGURE 1.11

An excerpt of ECG signal showing an episode of Sinus Bradycardia.

On the contrary, with a presence of variable SA node firing automaticity, the sinus node rate slow transiently, followed by atrial tachyarrhythmias, causes paroxysmal dizziness, confusion, and syncope due to the cardiac output reduction (Lilly, 2012). This phenomenon is called sick sinus syndrome (SSS).

Junctional and Ventricular Escape Rhythms

Considerable reduction of the SA node activity or the atrial conduction pathway contributes to the development of escape beats, originated in other pacemaker activity areas such as the AV node and bundle of His. These junctional escape beats modify the ECG characteristics, as follows (Lilly, 2012):

- Slower heart rate between 40 bpm and 60 bpm;
- Widened QRS complex due to the impulse conduction from areas not belonging to the rapidly conducting Purkinje fibers;
- Absence of a normal P-wave (the impulse is originated below the atria).

An excerpt of ECG signal showing an episode of junctional escape beat is presented in Fig. 1.12. This episode refers to the record 124 from MIT–BIH Arrhythmia Database (Goldberger et al., 2000).

Atrioventricular Node
First-Degree AV Block

The first-degree AV block is characterized by the difficulty of the atria impulses in reaching the ventricles, reflecting in the ECG a sustained delay prolongation between atrial and ventricular depolarization, a P–R interval higher than 0.2 s in all beats (Fig. 1.13). It can be caused by structural defects of the conduction pathway as seen, for example, in chronic conduction degenerative diseases and myocardial infarction (Lilly, 2012).

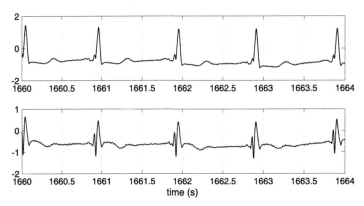

FIGURE 1.12

An excerpt of ECG signal showing an episode of Junctional Escape Beats.

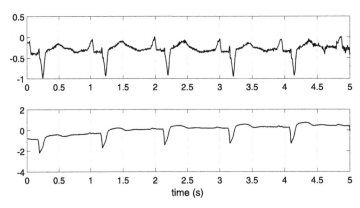

FIGURE 1.13

An excerpt of ECG signal showing an episode of First-degree AV block.

Second-Degree AV Block

In the second-degree AV block, the atria impulses fail to reach the ventricles in all beats. It is due a conduction failure from the AV node, where atria impulses propagates intermittently. The ECG of this intrinsic behavior is characterized by a QRS complex not always preceded by a P wave (Lilly, 2012). The second-degree AV block can have two different forms:

- Mobits Type I block is characterized by a gradual and progressive conduction defect between atria and ventricles until an impulse is completely blocked and ventricular stimulation lacks as a result of a single beat (Pastore et al., 2016; Lilly, 2012). The ECG shows a progressive P–R interval increment and R–R interval reduction between beats, until the next QRS complex is absent, starting the cycle anew. During the pause, the R–R interval is twice shorter than the previous,

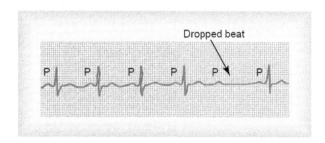

FIGURE 1.14

Example of a second degree AV block, showing a dropped beat (failure of the ventricles to receive the excitatory signals) (lead V3).

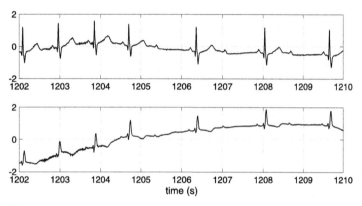

FIGURE 1.15

An excerpt of ECG signal showing an episode of Mobits Type-II block: register 231 from MIT–BIH Arrhythmia Database.

see Fig. 1.14 (this figure was published in Textbook of Medical Physiology, Arthur C. Guyton and John E. Hall, Chapter 13: Cardiac Arrhythmias and Their Electrocardiographic Interpretation, Page 149, Copyright Elsevier Inc. (2006)) (Pastore et al., 2009);

- Mobits Type II block is characterized by a sudden conduction interruption, where the AV node ceases to conduct two or more beats, unexpectedly, without any ECG warning. As consequence, the ECG shows sequential P waves without a correspondent QRS complex (Fig. 1.15). The His-Purkinje areas might play a role in this ceased behavior, resulting in abnormally wide QRS complexes and extensive infarction or chronic degenerative of the conduction pathway (Lilly, 2012).

Third-Degree AV Block

The third-degree AV block is characterized by a complete collapse of the impulses conduction between the atria and ventricles, dividing the heart into two unconnected

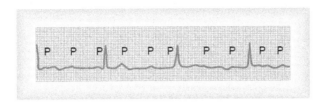

FIGURE 1.16

Example of a complete AV block (lead II).

zones, without P waves and QRS complexes sequential relationships (see Fig. 1.16; this figure was published in Textbook of Medical Physiology, Arthur C. Guyton and John E. Hall, Chapter 13: Cardiac Arrhythmias and Their Electrocardiographic Interpretation, Page 149, Copyright Elsevier Inc. (2006)). In this scenario, the atria depolarize by the SA node at a higher rate (Pastore et al., 2009), and the ventricles by their own intrinsic escape rate, usually between 40 bpm to 55 bpm (Lilly, 2012).

1.3.2 TACHYCARDIA

Sinus Tachycardia

Sinus Tachycardia is characterized by a SA node firing at rate above 100 bpm and normal P waves (positive at leads I, II, and aVF) and QRS complexes (Lilly, 2012; Pastore et al., 2016; Prystowsky and Klein, 1994), a consequence of a sympathetic increase and vagal stimulation decrease from the SA node. This arrhythmia is non-paroxysmal with graduated onset and offset. P-wave morphology can be slightly changed due to a shift in the site of the dominant pacemaker activity (Prystowsky and Klein, 1994). An excerpt of sinus tachycardia extracted from MIT–BIH Normal Sinus Rhythm Database (register 18177, PhysioBank) (Goldberger et al., 2000) is presented in Fig. 1.17.

Atrial Premature Beats

Atrial premature beats are commonly triggered by an ectopic atrium focus generating an abnormal wave shape preceding the expected P wave, and followed by normal QRS morphology complex, not positives in leads I, II, and aVF. If the ectopic focus trigger fires during the AV node refractory period, the abnormal P wave morphology is not followed by a QRS complex. However, with the ectopic focus trigger firing a bit delayed in the diastole, the impulse face parts of the His-Purkinje system in the refractory period, and the ventricules in a slower rate than normal, impacting in a significantly widened QRS complex. An excerpt of atrial premature contraction extracted from MIT–BIH Arrhythmia Database (register 100, PhysioBank) (Goldberger et al., 2000) is presented in Fig. 1.18.

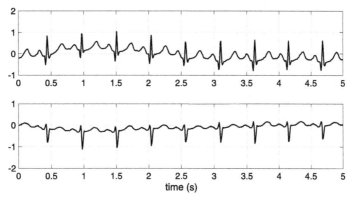

FIGURE 1.17

Excerpt of sinus tachycardia extracted from PhysioBank (MIT–BIH Normal Sinus Rhythm Database – register 18177).

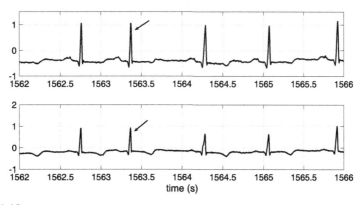

FIGURE 1.18

Excerpt of atrial premature beat extracted from PhysioBank (MIT–BIH Arrhythmia Database – register 100).

Atrial Flutter

On ECG, atrial flutter is characterized by a typical coarse "sawtooth" pattern of flutter waves, at rates of 300 bpm or above, and regularly undulating baseline, commonly best observed at leads II, III, and aVF (Lilly, 2012; Prystowsky and Klein, 1994; Pastore et al., 2016). Part of the flutter waves reach the AV node at the refractory period, slowing down the ventricular rate, usually one-half of the flutter rate. The arrhythmia mechanism is composed by a macroreentrant circuit defined as a circular reentry circuit, in which the wavefront propagates in a circuit that turn around an anatomic or functional substrate, likely to occur at the right atrium (Pastore et al., 2016). A common atrial flutter or counter-clockwise is characterized by a reentry

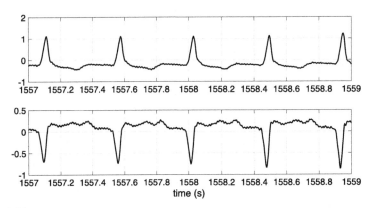

FIGURE 1.19

Excerpt of atrial flutter extracted from PhysioBank (MIT–BIH Arrhythmia Database – register 202).

with a wavefront propagating down to the right atrium anteriolateral wall, proceeding to the istmo cavotricuspid and going up to interatrial septum (Pastore et al., 2016).

In this scenario, flutter waves are:

- negative in leads II, III, and aVF;
- positive in lead V1;
- lower flutter waves amplitude at lead I and avL (Pastore et al., 2016).

Two types of atrial flutter are recognized (Prystowsky and Klein, 1994):

- Flutter Type 1: with rates between 250 bpm to 350 bpm and majority seen as negative flutter waves in the inferior limbs;
- Flutter Type 2: with higher rates between 350 to 450 bpm and predominantly positive deflection, as recorded in the inferior limbs, although can be converted to atrial fibrillation.

An excerpt of atrial flutter extracted from MIT–BIH Arrhythmia Database (register 202, PhysioBank) (Goldberger et al., 2000) is presented in Fig. 1.19.

Atrial Fibrillation

Atrial fibrillation is characterized by uncoordinated electrical atrial activation, at rates between 350 to 600 discharges/min, with consequent mechanical deterioration, associated fibrosis, and loss of atrial muscle mass, affecting the ability of the atria to pump blood effectively (NCC, 2006; Fuster et al., 2006). Since the atrial muscle fibres contract independently, there are no P-waves on the ECG, only irregular fibrillatory waves (f-waves) (Pastore et al., 2016), which might contribute to the formation of blood clots and thromboembolic events, increasing the risk of stroke five-fold (Wolf et al., 1996; Connolly, 2011). The f-waves have characteristics of low and variable amplitudes, and irregular baseline undulations (Prystowsky and Klein, 1994;

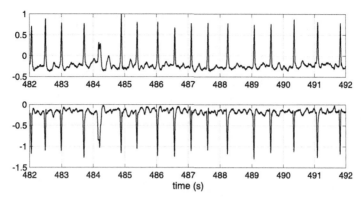

FIGURE 1.20

Excerpt of atrial fibrillation episode extracted from PhysioBank (MIT–BIH Arrhythmia Database – register 201).

Loures Salinet, 2013). The continuous wavefront activity reaching the AV node promotes irregular ventricular beats and consequently irregular contractions. The mean ventricular rate is higher than 100 bpm (Prystowsky and Klein, 1994; Lilly, 2012; Loures Salinet, 2013; Pastore et al., 2016). As the ventricular conduction system is normal, the QRS complex and T wave are of normal shape. However, detection of the T-wave end is a challenge in AF, given that T-wave and f-waves are issue of confusion (Salinet et al., 2013).

An excerpt of atrial fibrillation extracted from MIT–BIH Arrhythmia Database (register 201, PhysioBank) (Goldberger et al., 2000) is presented in Fig. 1.20.

Paroxysmal Supraventricular Tachycardias

Paroxysmal Supraventricular Tachycardias are characterized by atrial rates between 140 and 250 bpm, with two possible causal mechanisms. The first, a reentry loop observed within the AV node, SA node, or within the atria. The ECG lacks a P-wave due to the retrograde atrial depolarization occurring mostly simultaneously with the ventricular depolarization, where the P-wave may be "hidden" in the QRS complex (Lilly, 2012). Moreover, the ECG shows a regular tachycardia with the QRS complexes mostly narrowing, and the P wave is retrograded (inside or just after the QRS complex). These retrogrades P-waves can simulate a S wave at leads II, III, and aVF (pseudo S) and/or an R wave at V1 (pseudo R) (Pastore et al., 2016).

The second mechanism is due to increased automacity, and since it is mostly triggered by an ectopic atrial focus, the P-wave is upright, precedes the QRS, and is of abnormal shape (Lilly, 2012).

Atrial Tachycardia

This arrhythmia can be originated in atria, except in SA and AV nodes. It is characterized by an organized atrial electrical activity and the majority of the time it is possible to identify P-waves, followed by QRS complexes, with regular rate when AV conduc-

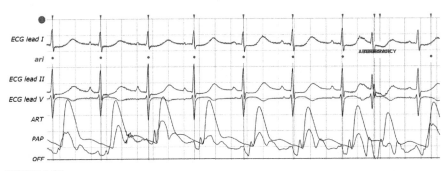

FIGURE 1.21

Excerpt of multifocal atrial tachycardia, extracted from PhysioBank MGH/MF Waveform Database – register mgh012.

tion is 1:1. In the ECG, P-waves and PR intervals are very similar. P-waves are negative in leads I, II, and aVF, with a P rate of 100 bpm and 250 bpm (Pastore et al., 2016; Prystowsky and Klein, 1994). If AV conduction is different than 1:1, the R–R interval can be either regular or irregular; there is also a systematic PR interval increment followed by an AV block pattern (P-wave is blocked) (Pastore et al., 2016).

Multifocal Atrial Tachycardia

Multifocal Atrial Tachycardia is characterized by multiple ectopic atrial focus, reflecting an atria disorganized electrical activity, between 160 bpm and 250 bpm, but of lower degree when compared with atrial fibrillation (Pastore et al., 2016; Lilly, 2012). As consequence, P-wave is of different morphology over beats, and the beating pattern persists for three P-waves or more, each beat preceding the respective QRS complexes (Pastore et al., 2016) (see Fig. 1.21). Moreover, PR interval also differs between beats, and AV conduction rate is commonly of 1:1 ratio (Pastore et al., 2016; Lilly, 2012).

Wolff–Parkinson–White Syndrome

This arrhythmia presents an abnormal pathway conduction connection tract between atria and ventricles, allowing the atria wavefront to travel to the ventricles through the AV node and the accessory pathway. Clinically, the most common pathway identified is the bundle of Kent, and it can occur in both anterograde and retrograde directions. The abnormal pathway tract bypasses the AV node, stimulating the ventricles earlier giving rise to two different ventricular patterns (Lilly, 2012):

- PR interval is shortened (no delay prior the ventricular stimulation);
- a slurred rather than sharper upstroke of the QRS (termed "delta wave"), with an abnormally widened QRS complex, due to fusion of the impulses arriving at the ventricles via two separate routes.

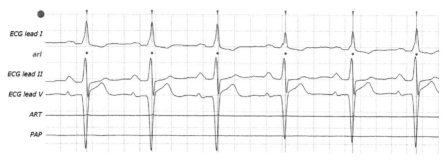

FIGURE 1.22

Excerpt of ECG signal presenting Wolff–Parkinson–White pattern syndrome, extracted from PhysioBank MGH/MF Waveform Database – register mgh037.

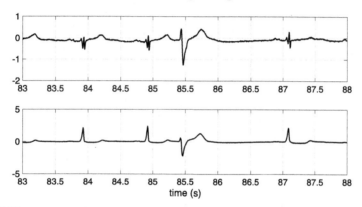

FIGURE 1.23

Excerpt of ECG signal presenting a ventricular premature beat, extracted from PhysioBank MIT–BIH Arrhythmia Database – register 114.

An excerpt of ECG signal presenting Wolff–Parkinson–White pattern syndrome, extracted from PhysioBank MGH/MF Waveform Database – register mgh037, is illustrated in Fig. 1.22.

Ventricular Premature Beats

Ventricular premature beats arise when an ectopic ventricular focus triggers an action potential. The ECG shows a widened QRS complex, since the ectopic impulse propagates through the ventricles via slow cell-to-cell rather than the normal rapid conduction system pathway (see Fig. 1.23). The ectopic beat is not related to a preceding P wave, and this arrhytmia is often asymptomatic and benign, and also commonly seen in healthy individuals (Lilly, 2012).

Ventricular Tachycardia

It is characterized by three or more sequential ventricular premature beats and can be divided into sustained ventricular tachycardia (with episodes lasting for 30 s or

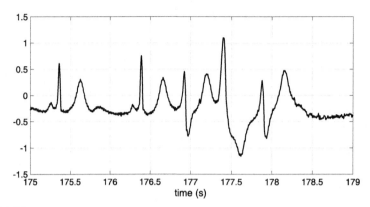

FIGURE 1.24

Excerpt of ECG signal presenting an episode of ventricular tachycardia lasting 2-s, extracted from PhysioBank MIT–BIH Arrhythmia Database – register 106.

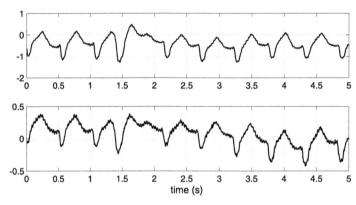

FIGURE 1.25

Excerpt of ventricular fibrillation, extracted from PhysioBank MIT–BIH Malignant Ventricular Arrhythmia Database – register 418.

longer) and nonsustained ventricular tachycardia. In the ECG, the QRS complexes are abnormally broad with a monomorphic or variable shapes, suggesting a presence of multiple ventricular foci (Fig. 1.24). Symptomatic or sustained ventricular episodes can lead to ventricular fibrillation (Lilly, 2012).

Ventricular Fibrillation

Ventricular Fibrillation is the most life-threatening arrhythmia in clinical practice that results in a severe cardiac output drop, leading to cardiac arrest and sudden death, if not promptly terminated. It is characterized by disorganized high-frequency activity from the ventricles seen on the ECG by waves varying in amplitude and morphology without discrete QRS complexes (Fig. 1.25). This arrhythmia is often related to patients with severe heart disease (Lilly, 2012).

1.4 CONCLUSIONS

In this chapter, we detailed physiological aspects of electrical activity of the heart, the basic concepts related to the electrocardiogram (ECG) signal, and how the ECG should be interpreted for understanding and detecting heart rhythms and conduction disorders. It is obvious that the ECG signal provides important information regarding heart activity, and important events may be diagnosed through the relationships among the characteristic waves, amplitudes, intervals, morphologies, and occurrence or absence of waveforms. It is important to emphasize that some adverse events may be predicted by analyzing the evolution of some metrics derived from the ECG signal processing, such as heart rate variability. Knowing the significant potential of the ECG information for providing useful diagnostic insights motivates research concerned with its accurate preprocessing and feature extraction.

REFERENCES

Connolly, S.J., 2011. Atrial fibrillation in 2010: advances in treatment and management. Nature Reviews Cardiology 8 (2). nrcardio-2010.

Fuster, V., Rydén, L.E., Cannom, D.S., Crijns, H.J., Ivanov, P.C., Curtis, A.B., Ellenbogen, K.A., Halperin, J.L., Le Heuzey, J.Y., Kay, G.N., 2006. ACC/AHA/ESC 2006 guidelines for the management of patients with atrial fibrillation-executive summary: a report of the American College of Cardiology/American Heart Association Task Force on practice guidelines and the European Society of Cardiology Committee for Practice Guidelines (Writing Committee to Revise the 2001 Guidelines for the Management of Patients with Atrial Fibrillation) Developed in collaboration with the European Heart Rhythm Association and the Heart Rhythm Society. European Heart Journal 27, 1979–2030.

Gari, D.C., Francisco, A., Patrick, E., 2006. Advanced Methods and Tools for ECG Data Analysis. Artech House, Inc.

Goldberger, A.L., Amaral, L.A., Glass, L., Hausdorff, J.M., Ivanov, P.C., Mark, R.G., Mietus, J.E., Moody, G.B., Peng, C.K., Stanley, H.E., 2000. PhysioBank, PhysioToolkit, and PhysioNet. Circulation 101 (23), e215–e220.

Hall, J.E., 2011. Guyton and Hall Textbook of Medical Physiology. Saunders Elsevier, Philadelphia.

Lilly, L.S., 2012. Pathophysiology of Heart Disease: a Collaborative Project of Medical Students and Faculty. Lippincott Williams & Wilkins.

Loures Salinet Jr., J., 2013. High Density Frequency Mapping of Human Intracardiac Persistent Atrial Fibrillation Electrograms. PhD thesis. University of Leicester.

Luz, E.J.d.S., Schwartz, W.R., Cámara-Chávez, G., Menotti, D., 2016. Ecg-based heartbeat classification for arrhythmia detection: a survey. Computer Methods and Programs in Biomedicine 127, 144–164.

National Collaborating Centre (NCC) for Chronic Conditions, 2006. Atrial Fibrillation: National Clinical Guideline for Management in Primary and Secondary Care. Royal College of Physicians, London.

Pastore, C., Pinho, C., Germiniani, H., Samesima, N., Mano, R., 2009. Diretrizes da sociedade brasileira de cardiologia sobre análise e emissão de laudos eletrocardiográficos. Arquivos Brasileiros de Cardiologia 93 (3), 1–19.

Pastore, C.A., Samesima, N., Tobias, N., Pereira Filho, H.G., 2016. Eletrocardiografia atual: curso de serviço de eletrocardiografia do InCor. Atheneu.

Prystowsky, E.N., Klein, G.J., 1994. Cardiac Arrhythmias: an Integrated Approach for the Clinician. McGraw–Hill.

Salinet Jr., J.L., Madeiro, J.P., Cortez, P.C., Stafford, P.J., Ng, G.A., Schlindwein, F.S., 2013. Analysis of QRS-T subtraction in unipolar atrial fibrillation electrograms. Medical & Biological Engineering & Computing 51 (12), 1381–1391.

da Silva, HP, Carreiras, C., Lourenço, A., Fred, A., das Neves, R.C., Ferreira, R., 2015. Off-the-person electrocardiography: performance assessment and clinical correlation. Health and Technology 4 (4), 309–318.

Wasilewski, J., Poloński, L., 2012. An introduction to ECG interpretation. In: ECG Signal Processing, Classification and Interpretation. Springer, pp. 1–20.

Wolf, P.A., Benhamin, E.J., Belanger, A.J., Kannel, W.B., Levy, D., D'agostino, R.B., 1996. Secular trends in the prevalence of atrial fibrillation: the framingham study. American Heart Journal 131 (4), 790–795.

ECG Signal Acquisition Systems

João Loures Salinet Jr., Olavo Luppi Silva
*Center of Engineering, Modeling and Applied Social Sciences,
Federal University of ABC, São Paulo, Brazil*

2.1 INTRODUCTION

The ECG acquisition systems, like many other biomedical devices, are designed to measure signals from the human body, aiming to help clinicians at the diagnoses and treatment of diseases. Before start discussing the design of an ECG system circuitries, projectists need to know the intrinsic characteristics of the ECG and the possible factors which could affect it (Webster, 2006a).

2.2 ELECTRODE-PATIENT INTERFACE

The currents and charges generated by the heart electrical activity are propagated inside the patient's body to the torso by ions, where devices for conversion to electronic ECG are needed. This conversion, ions to electrons, occurs at the electrode-patient interface by an underlying mechanism identified as charge-transfer. Characteristics from both electrode-patient interface and skin under the electrode are directly involved on the charge-transfer mechanism, affecting conversion of the resultant electronic heart currents and consequently the ECG (Webster, 2006b).

In the ECG electrodes, a high electrical performance can generally be obtained by non-noble materials, for example, silver–silver chloride (Ag–AgCl), the most commonly used material for ECG electrodes. When this metallic electrode is in contact with an electrolyte (skin or electrode gel), an electrochemical reaction occurs by ion-exchange (Eq. (2.1)), based on the tendency of the metal atoms (M) to lose electrons, and the metal ions (M^{+n}) are moved to the electrolyte, implicating an electrode that is negatively charged when compared with the electrolyte (oxidation reaction). Similarly, in the electrolyte, the metal ions (M^{+n}) take the n electrons to form metal atoms (M) that are deposited onto the electrode, implicating an electrode that is positively charged with respect to the electrolyte (reduction reaction) (Webster, 2006b):

$$M \Leftrightarrow M^{+n} + n^{e-} \tag{2.1}$$

Developments and Applications for ECG Signal Processing. https://doi.org/10.1016/B978-0-12-814035-2.00008-6

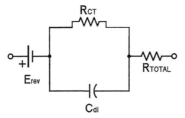

FIGURE 2.1

Electrode-electrolyte interface equivalent circuit. C_{dl} represents the double-layer capacitance; R_{CT}, the charge transfer resistance; R_{TOTAL}, lead and electrode resistances, and E_{rev}, the equilibrium potential.

The ion-exchange rates into both directions are balanced at equilibrium conditions, leading to a resultant current equal to zero (current flowing in opposite directions are equal). However, the resulting potential difference at the electrode-electrolyte interface, termed as half-cell potential, is nonzero and depends on the concentrations of both ions and metal electrode and can obtained by the Nerst Equation (Eq. (2.2)) (Webster, 2006b).

$$E_{half-cell} = E_0 + \frac{RT}{nF} \times \log \frac{\text{Concentration of oxidation form}}{\text{Concentration of reduced form}}. \quad (2.2)$$

The parameter $E_{half-cell}$ is the half-cell potential, E_0 is the standard half-cell potential measured relative to the standard hydrogen electrode, R is the universal gas constant, n is number of electrons involved in the reaction and T is the absolute temperature in Kelvin. Concentration is used here by assuming that the solution is infinitely dilute (Webster, 2006b).

Since an ECG signal is obtained when two equal electrodes are placed at the torso, it can be assumed that the electrolyte/electrode interface are of the same properties, resulting in equals half-cell potentials where its equivalence is of zero voltage offset at the input of the differential amplifier, not affecting the ECG signal. In practice, small differences occur implicating in a voltage offset, which can sometimes change overtime affecting the measured ECG.

At the electrode-electrolyte interface, positive ions are attracted by the electrode (negatively charged due the oxidation reaction) and negative ions by the electrolyte (positively charged due the reduction reaction) creating a double layer capacitance at the interface (C_{dl}). This capacitance is connected in parallel to the charge-transfer resistance (R_{CT}), where transference of ions charged at the torso, coming from the heart electrical activity, to electrons for the ECG acquisition system occurs. At the double layer capacitance, exchange currents across its parallel plates on both directions face the charge transfer resistance which contributes to electrons conversion. A simplified equivalent electrical circuit model from the electrode-electrolyte interface is presented in Fig. 2.1.

At the interface, some dc (or faradaic) current can leak across the double layer due to the undergoing interface electrochemical reactions (Eq. (2.2) and C_{dl}). These reactions experience a charge transfer resistance, and can be thought as shunting the nonfaradaic, double-layer capacitance. The charge transfer resistance equation is shown in Eq. (2.3), where i_0 is the current flowing across the interface in both directions (no flows under equilibrium):

$$R_{CT} = \frac{RT}{nF} \times \frac{1}{i_0}. \tag{2.3}$$

A good electrode system will therefore be characterized by a large value of exchange current or a low value of R_{CT}, resulting in an unimpeded charge transference and consequently a small voltage drop will be present across the interface (Webster, 2006b).

2.3 LEADS SYSTEM

The ECG signals are electrical representation from the cardiac electrical activity, noninvasively captured by two or more electrodes. It results that the acquired ECG signal is the difference between the two electrodes (one positive electrode and other the negative or reference electrode), which comprise a single ECG lead. There are different types of electrodes that could be used to acquire ECG signals, and the most common are superficial noninvasive electrodes, with a conductive gel between electrode metallic part and patient skin, to reduce the skin-electrode impedance (Gari et al., 2006).

The number of electrodes to be placed at predetermined areas of the torso would allow projection of the electrical heart activity in different electrical axis, providing a robust interpretation of the undergoing heart electrical activity and allowing the clinicians to diagnose heart disorders through the spatial correlation of ECG events on specific leads. A variety of leads configurations have been proposed in literature (Macfarlane et al., 2010a), and the choice of lead configuration should take into account the type of cardiac activity that a patient is expected to experience (Gari et al., 2006).

2.3.1 ECG MONITORING HOSPITAL LEADS

As above described, two electrodes comprise a single ECG lead to represent the electrical heart activity in a differential electrical axis. The use of additional leads allows distinct spatial perspective of the electrical activity of the heart and has shown to improve the signal-to-noise ratio. That is, more electrodes in similar locations would allow clinicians to improve the chances of identification of clinical features (Gari et al., 2006).

Some bedside cardiac monitors are capable of recording ECG just by a single exploring electrode which is on chest (precordial lead), where lead V_1 (see

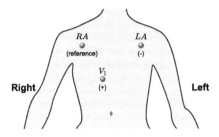

FIGURE 2.2

Lead placement for Modified Chest Lead (MCL) configuration using a 3-lead system. The positive electrode is placed in V_1 and the negative electrode at the left infraclavicular fossa. The reference (ground) electrode, here shown in the right arm, could be placed anywhere on the torso. RA: right arm; LA: left arm.

Fig. 2.2) proved to be of high sensitivity for cardiac arrhythmias detection (Gari et al., 2006). Moreover, wrist and ankle electrodes are suggested to be placed on the torso (Macfarlane et al., 2010a). Simple 3-electrode bipolar system used on hospital monitoring standards basically applies the Modified Chest Lead (MCL) configuration, where the positive electrode is placed in V_1, the negative electrode at the left infraclavicular fossa, and the reference electrode on the right arm position, but free to be placed anywhere on the torso (Fig. 2.2). Many other lead systems such as reduced leads, orthogonal leads, and derived 12-leads are also be used in hospital ECG monitoring (Macfarlane et al., 2010a) and will be introduced at the following sections.

2.3.2 STANDARD LEADS NOMENCLATURE

Before describing in detail the standard 12-lead ECG, the leads nomenclature and definition are firstly introduced. The standard leads nomenclature of the 12-lead ECG is divided into bipolar limb leads, unipolar chest leads and augmented unipolar limb leads.

A bipolar limb lead, by definition, measures the potential difference between two limbs, in which the most commonly used are the introduced by (Einthoven et al., 1950), comprising the potential difference between the left and right arms (called lead I), left leg and right or left arms (called lead II and III respectively) Fig. 2.3. The unipolar chest leads is part of the humans thoracic lead system, commonly used at clinical practice by six precordial leads. Each of them measures the potential difference between a positive exploratory electrode (placed on specific position on the chest and are called as V_1 to V_6) and the Wilson Central Terminal reference (see next section and Fig. 2.5 for detailed explanation). At last, the augmented unipolar limb leads measures the potential of one limb (positive electrode) referenced against a combination of the other limb electrodes (disregard right leg). The positive electrodes for these augmented leads are located on the right arm (called of aVR), the left arm (aVL), and the left leg (aVF). In practice, the exploratory positive electrodes used in the augmented unipolar limb leads are the same electrodes used for bipolar limb leads (Fig. 2.6). At the next sections, each of the standard 12-lead ECG leads, is described in details.

2.3.3 THE 12-LEAD ECG

The standard 12-lead ECG is a noninvasive representation of the electrical activity of the heart, where the thorax is assumed to be a homogeneous conductor, depicted by the difference between electrodes on predetermined body location (leads), including the three bipolar limb leads (I, II and III), six unipolar chest leads (V_1 to V_6), and three augmented unipolar limb leads (aVR, aVL and aVF).

The three bipolar limb leads introduced by Einthoven et al. (1950) depict the potential difference between the left arm (E_{LA}) and right arm (E_{RA}) denoted at lead I (Eq. (2.4)); left leg (E_{LL}) and right arm denoted as lead II (Eq. (2.5)), and left leg and left arm denoted as lead III (Eq. (2.6)). The relation between the three bipolar limb leads, known as Einthoven's law, is shown at (Eq. (2.7)), obtained through the mathematical manipulation of the above bipolar limb leads equations. Fig. 2.3 shows each bipolar limb represented by its respective limb vector, where the relation between them forms a triangle, known as Einthoven triangle. The right leg electrode is required for reducing the common-mode interference and could be placed elsewhere in the body, however, it is placed at this limb by convenience.

$$I = E_{LA} - E_{RA} \qquad (2.4)$$
$$II = E_{LL} - E_{RA} \qquad (2.5)$$
$$III = E_{LL} - E_{LA} \qquad (2.6)$$
$$II = I + III \qquad (2.7)$$

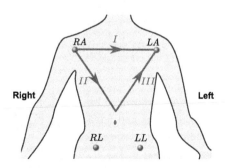

FIGURE 2.3

Triangular representation of bipolar limb leads I, II, and III known as Triangle of Einthoven. RA: right arm; LA: left arm; LL: left leg; RL: right leg.

The deflections observed on a ECG signal are seen as a result of the equivalent cardiac wavefront propagating (approximating or moving away) across a chosen unipolar or bipolar electrode. By convention, if the cardiac wavefront propagates to the limb direction, the respective limb bipolar ECG will show a positive deflection. Otherwise, if the cardiac wavefront moves into opposite limb vector direction, by convention, the respective limb bipolar ECG will show a negative deflection (please see Fig. 2.4). Extrapolating, a certain cardiac wavefront propagating into lead I direction will therefore show a positive deflection for lead I and a negative deflection for lead III. For the unipolar electrodes, the deflections are also under the same behavior.

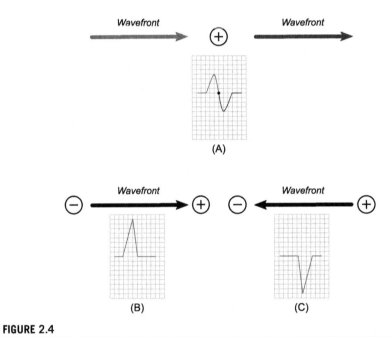

FIGURE 2.4

The bipolar and unipolar leads axes and the current flow vector at the heart in a certain instant of time. The polarity of the QRS deflection is obtained due to the direction of the current vector relative to the axis of the chosen lead.

The Wilson Central Terminal, shown in Fig. 2.5A, is defined as the sum of the potentials of the right and left arms and left leg (Eq. (2.8)), and it has been used as the reference to the unipolar recordings electrodes on torso, such the unipolar limb leads (V_1 to V_6) (depicted in Fig. 2.5B). Mathematically, any of the six unipolar chest leads display the potential heart activation on areas close to the heart on the torso projection (E_{1-6}) referenced to the Wilson Central Terminal (WCT, Eq. (2.9)):

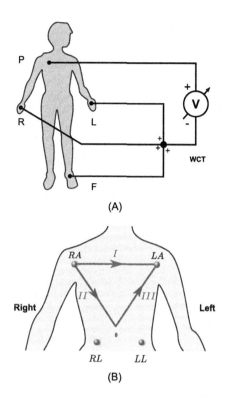

(A)

(B)

FIGURE 2.5

(A) Unipolar limb lead circuit obtained by the difference between the potential of a single P electrode at the chest and the Wilson Central Terminal; (B) torso representation with both bipolar limb leads (I, II, and III) and unipolar chest leads V_{1-6} (from Macfarlane et al., 2010a).

$$WCT = \frac{1}{3} \times ((E_{RA}) + (E_{LA}) + (E_{LL}));$$ (2.8)

$$V_i = E_i - WCT, \quad i = 1\ldots6.$$ (2.9)

where V_1 and V_2 are localized at the right and left-hand side at the fourth intercostal space, V_3 between V_2 and V_4, V_4 at fifth intercostal, V_5 between V_3 and V_6, V_6 at the midaxillary line.

The remaining three leads that compose the 12 lead ECG, called augmented unipolar limb leads (aVR, aVL and aVF) (Goldberger, 1942), have the positive electrode in each of the limbs (right atrium, or left atrium or left leg). The reference for each of them is a modified version of the Wilson Central Terminal (called of GT), where the other two limbs is then averaged (see Fig. 2.6A and Eqs. (2.10) to (2.12)). It allows increasing the voltages amplitude ("augmented"), measured by the unipolar limbs, and the letter "a" is added to represent the increased voltage amplitudes at the

ECGs.

$$aVR = E_{RA} - GT = E_{RA} - \frac{1}{2}(E_{LA} + E_{LL}) \qquad (2.10)$$

$$aVL = E_{LA} - GT = E_{LA} - \frac{1}{2}(E_{RA} + E_{LL}) \qquad (2.11)$$

$$aVF = E_{LL} - GT = E_{LL} - \frac{1}{2}(E_{RA} + E_{LA}) \qquad (2.12)$$

Deriving Eq. (2.10), as presented below, the equivalent right arm potential referenced to GT gives a potential of higher amplitude (factor of 1.5x) when compared to its right arm unipolar potential (E_{RA}).

$$aVR = E_{RA} - GT \quad =$$
$$E_{RA} - \frac{1}{2}(E_{LA} + E_{LL}) \quad =$$
$$\left(\frac{2}{2} + \frac{1}{2}\right)E_{RA} - \frac{1}{2}(E_{RA} + E_{LA} + E_{LL}) \quad =$$
$$\frac{3}{2}E_{RA} - \frac{1}{3}(E_{RA} + E_{LA} + E_{LL}) \quad =$$
$$\frac{3}{2}E_{RA} - WCT$$

The augmented unipolar limb aVR can also be calculated as a function of the ECG bipolar limb leads. For this, we need to consider Eqs. (2.4) to (2.7).

$$aVR = E_{RA} - \frac{1}{2}(E_{LA} + E_{LL}) \quad =$$
$$\frac{1}{2}(E_{RA} - E_{LA}) + \frac{1}{2}(E_{RA} - E_{LL}) \quad =$$
$$\frac{1}{2}(-I) + \frac{1}{2}(-II) \quad =$$
$$-\frac{1}{2}(I + II)$$

Reciprocally, aVL and aVF are also obtained as a linear combination of two out of the three bipolar limb leads, see below:

$$aVL = E_{LA} - \frac{1}{2}(E_{RA} + E_{LL}) \quad =$$
$$\frac{1}{2}(E_{LA} - E_{RA}) + \frac{1}{2}(E_{LA} - E_{LL}) \quad =$$
$$\frac{1}{2}(I) + \frac{1}{2}(-III) \quad =$$
$$\frac{1}{2}(I - III)$$

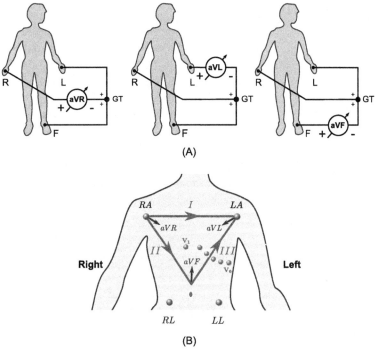

(A)

(B)

FIGURE 2.6

(A) Schematic of augmented unipolar limb lead circuit, in this example avR, obtained by the difference between the potential of a single R electrode at the right arm and the GT (adapted from Macfarlane et al., 2010a). (B) Torso representation with the 12-lead ECG.

$$aVF = E_{LL} - \frac{1}{2}(E_{RA} + E_{LA}) \quad =$$

$$\frac{1}{2}(E_{LL} - E_{RA}) + \frac{1}{2}(E_{LL} - E_{LA}) \quad =$$

$$\frac{1}{2}(II) + \frac{1}{2}(III) \quad =$$

$$\frac{1}{2}(II + III)$$

Since the augmented unipolar limb leads are a linear combination of the bipolar limb leads, and according to Einthoven's triangle equation (Eq. (2.7)), any bipolar limb lead is calculated as a function of the other two ones, it is possible to conclude that with only two bipolar limbs, the third one and the augmented unipolar limb leads are obtained by a simple mathematical manipulation. It is of significant importance to computer-based electrocardiography systems and memory saving. The illustration of all leads contemplating the standard 12-lead ECG is now summarized in Fig. 2.6B.

2.3.4 FRANK SYSTEM

The equivalent cardiac activity at a certain time moment within the cardiac cycle can be represented mathematically by a vector of specific magnitude and direction represented spatially in a three-dimensional orthogonal axes: laterolateral (right-to-left or X axis), craniocaudal (head-to-feet or Y axis), and antero-posterior (front-to-back or Z axis), see Fig. 2.7. The Vectocardiogram (VCG) is a variante of electrocardiography and records the same heart's events, but now the heart's electrical vector is presented a propagation of the heart in the form of loops projected on three planes. Each plane is obtained by a combination of two axes, with the frontal plane by the combination of both Y and Z axes, transverse by the X and Y axes and sagittal by X and Z axes. This method has helped to supplement information not easily detectable through traditional electrocardiographic analysis (Macfarlane et al., 2010a).

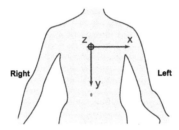

FIGURE 2.7

The three orthogonal axes of the Vectocardiogram. X: laterolateral (or right-to-left); Y: craniocaudal (head-to-feet); Z: antero-posterior (front-to-back). Three planes are created from these axes, with the frontal plane by the combination of both Y and Z axes, transverse by the X and Y axes and sagittal by X and Z axes.

From the several orthogonal corrected leads systems proposed, the Frank's method (Frank, 1956) has been the most popular protocol accepted by the scientific community, where 7 electrodes (A, C, E, I, M, H, and F, see Fig. 2.8) are used to determine X, Y, and Z components. These components, combined two-by-two, originate three orthogonal planes, where spatial curves representative of the electrical heart's phenomenon are projected.

In this method, seven electrodes are strategically placed at predetermined thorax locations (Fig. 2.8), allowing the representation of electrical cardiac vector components (Frank, 1956) at the planes. The axis X is derived from the electrodes A, C, and I; Y axis is derived from electrodes H, M, and F; and Z axis is derived from electrodes A, C, E, I, and M. The vectocardiograms are created based on Frank's lead system, where three orthogonal leads corresponding to the axes of the body (V_X, V_Y, and V_Z) are calculated (Eqs. (2.13) to (2.15)). The coefficients below were computed from the Frank-lead system to derive the three orthogonal leads (V_X, V_Y, and V_Z). ECG leads V_3 and V_6 are common with electrodes C and A from the VCG, respectively

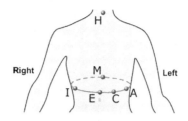

FIGURE 2.8

Electrodes position at the Frank modified lead system. Anatomical description: A) left midaxillary line, E) midsternum line, C) half way between A and E, I) right midaxillary line, M) midspine position, H) back of the neck, F) left foot. RL is the right leg electrode, commonly used also on the 12-lead ECG, as ground (Macfarlane et al., 2010a).

(Macfarlane et al., 2010a).

$$V_X = 0.61V_A + 0.171_C - 0.781V_I \qquad (2.13)$$

$$V_Y = 0.655V_F + 0.345V_M - 1.0V_H \qquad (2.14)$$

$$V_Z = 0.133V_A + 0.736V_M - 0.264V_I - 0.374V_E - 0.231V_C \qquad (2.15)$$

In cases where the electrodes are not positioned according to the Frank modified lead system, the orthogonal V_X, V_Y, and V_Z leads can be obtained through the standard ECG leads, by a weighted sum between independent leads, Eqs. (2.16) to (2.18).

A summary of the all transfer coefficients derived from the 12-lead ECG through the inverse Dower transformation matrix is presented at Table 2.1 (Edenbrandt and Pahlm, 1988):

$$V_X = -0.172V_1 - 0.074V_2 + 0.122V_3 + 0.231V_4 + 0.239V_5$$
$$+ 0.194V_6 + 0.156I - 0.011I; \qquad (2.16)$$

$$V_Y = +0.057V_1 - 0.019V_2 - 0.106V_3 - 0.022V_4 + 0.041V_5$$
$$+ 0.048V_6 - 0.227I + 0.887II; \qquad (2.17)$$

$$V_Z = -0.229V_1 - 0.31V_2 - 0.246V_3 + 0.063V_4 + 0.055V_5$$
$$+ 0.108V_6 + 0.022I + 0.102II. \qquad (2.18)$$

Transfer coefficients proposed by two other derived transformation matrices methods are shown in Tables 2.2 and 2.3 (Uijen et al., 1988; Willems, 1984). Observe that they also differ at the leads positions to calculate the orthogonal V_X, V_Y, and V_Z leads.

Table 2.1 Transfer coefficients XYZ derived from 12-lead ECG, proposed by (Edenbrandt and Pahlm, 1988)

	V_1	V_2	V_3	V_4	V_5	V_6	I	II
V_X	−0.172	−0.074	0.122	0.231	0.239	0.194	0.156	−0.01
V_Y	0.057	−0.019	−0.106	−0.022	0.041	0.048	−0.227	0.887
V_Z	−0.229	−0.31	−0.246	0.063	0.055	0.108	0.022	0.102

Table 2.2 Transfer coefficients XYZ derived from 12-lead ECG, proposed by (Uijen et al., 1988)

	RA	LA	V_1	V_2	V_3	V_4	V_5	V_6
V_X	−0.52	0.82	−0.01	0.04	0.04	0.05	0.07	0.37
V_Y	−1.53	−1.09	0.03	−0.02	−0.02	0.03	−0.07	0.08
V_Z	0.43	−0.01	−0.26	−0.28	−0.14	0.04	−0.15	0.34

Table 2.3 Transfer coefficients XYZ derived from 12-lead ECG, proposed by (Willems, 1984)

	II	III	V_1	V_2	V_3	V_4	V_5	V_6
V_X	0.58	−0.82	−1.27	−0.55	0.72	1.86	1.92	1.53
V_Y	2.58	3.04	−0.71	−0.71	0.1	0.35	0.12	−0.15
V_Z	−0.8	−1.62	−1.71	−2.26	−2.02	−0.8	0.31	0.97

Scaling:

V_X (measured) $= 0.1340 V_X$ (calculated);
V_Y (measured) $= 0.1565 V_Y$ (calculated);
V_Z (measured) $= 0.1220 V_Z$ (calculated).

2.3.5 EXERCISE ECG LEAD SYSTEMS

Different bipolar chest leads were proposed in the past for evaluating electrocardiography exercise testing. Currently, due the availability of computer-based exercise ECG analysis methods, the standard 12-lead ECG is the most common used in clinical practice. Exercise testing has been nowadays evaluated based on a modified version of the conventional 12-lead ECG, with the exercise performed on a treadmill or a bicycle ergometer. Within the differences, it can be highlighted that the right arm electrode is moved to below the clavicle in a point at the infraclavicular fossa medial, with the left arm at the corresponding position for the left side. The left leg electrode is recommended to be placed on the left iliac crest and the right leg electrode in the region of the right iliac fossa (Macfarlane et al., 2010a).

2.3.6 BODY SURFACE MAPPING LEAD SYSTEMS

Several efforts have been made to improve the understanding of heart disorders mechanisms, but the reduced number of electrodes (i.e. low spatial resolution) is still the major limitation for precise diagnosis of some heart diseases (Macfarlane

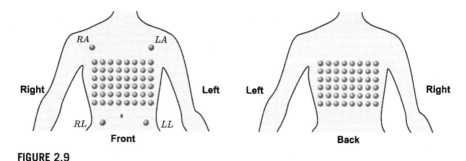

RA   LA

Right   Left

RL   LL

Front

Left   Right

Back

FIGURE 2.9

The electrode placement of Body Surface Potential Mapping system (BSPM) with 100 electrodes.

et al., 2010b; Taccardi et al., 1998; van der Graaf et al., 2014). van der Graaf et al. (2014) argues that a device containing a high spatial resolution of simultaneous noninvasive recordings would allow better detection of local variations in heart activity, helping in the diagnosis of heart diseases. A device called Body Surface Potential Mapping (BSPM) has been developed and utilized in both research and clinical practices (Ramanathan et al., 2004), allowing real-time acquisition of body surface potentials over a range of 32 to 300 electrodes, providing better diagnostics of several heart disturbances (Macfarlane et al., 2010b; Taccardi et al., 1998; van der Graaf et al., 2014). Fig. 2.9 shows an example of a BSPM electrodes layout with 100 electrodes (52 electrodes on front and 48 on the back).

In addition to these advantages, BSPM allows investigators to map patterns of propagation through a three-dimensional color-coded torso map of the electrical activity (Van Oosterom and Oostendorp, 2004) under abnormal heart rhythms, and inferences can be drawn with the normal patterns obtained from healthy individuals (Macfarlane et al., 2010b). Another BSPM system advantage is the possibility to estimate potentials at the heart epicardium by means of a mathematical inversion solution, based on the high density body surface potentials (Macfarlane et al., 2010b; Ramanathan et al., 2004; Van Oosterom and Oostendorp, 2004).

The ECG acquisition and conditioning systems described above systems can be done by conventional circuitry. In the following sections, the principal analog circuits that can be used to acquire ECG signals are described in detail.

2.4 ELECTRIC DESIGN GUIDELINES

An ECG lead system instrumentation is designed to measure subtle differences of potential at skin as a result of heart's depolarization wave. As can be seen in Fig. 2.10, it is difficult to measure only the biopotential of electrocardiogram. What is actually measured is a sum of the desired ECG with several other signals such as biopotential activity of muscles and nerves beneath electrodes, 60 Hz/50 Hz power line interference, low frequency motion artifacts, DC electrode contact potentials (half-cell potential), and all sort of high-frequency noise produced by electromagnetic waves.

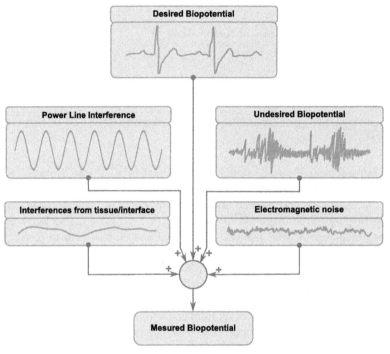

FIGURE 2.10

Components of measured biopotential.

Heart cell resting potentials are of the order of tenths of millivolts, but the attenuation produced by muscles and skin tissue makes ECG signal available at the electrode to the order of only a few millivolts. However, the sum of other undesired potentials are oftentimes one order of magnitude (or more) larger than that. That is why we need a signal conditioner circuit between the electrode and the digitalization stage. The signal conditioner is a set of analog circuits that manipulates an analog signal to meet the requirements of the next stage, which, in this context, is the analog for digital conversion. Additionally, the signal conditioner should accomplish other design requirements:

- amplify ECG signal without distortion;
- reject external interferences;
- filter artifacts produced by human body;
- protect patient and operator from electrical shock hazard;
- avoid damage by overvoltage produced by defibrillators and electrosurgical equipment.

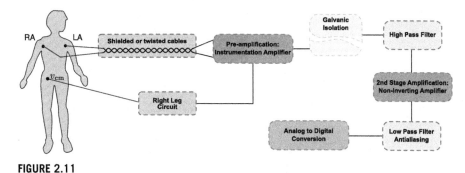

FIGURE 2.11

Main components of ECG signal conditioner.

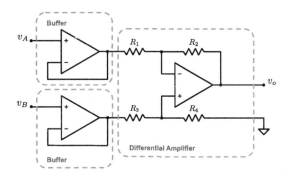

FIGURE 2.12

Differential amplifier with voltage followers for increased input impedance.

A classical topology of an ECG signal conditioner is composed by several sub-circuits, each one responsible for different operations shown in the block diagram of Fig. 2.11.

2.4.1 ACQUISITION AND AMPLIFICATION

The first and most important part of signal conditioner is the pre-amplification stage. Since an ECG derivation is the potential difference between two leads, the naive solution would be to use a difference amplifier. To increase input impedance, two voltage followers (buffers) are placed at the input as shown in Fig. 2.12.

An ideal op-amp is an amplifier with infinite open-loop gain, infinite input resistance, and zero output resistance. The consequence of these assumptions is that the inputs draw no current ($i_+ = 0$, $i_- = 0$) and the op-amp attempts to do whatever is necessary to make the voltage difference between the inputs equal to zero ($v_+ = v_-$) (Horowitz and Hill, 1989). Considering the ideal amp-op model, the relation between the output v_o and the inputs v_A and v_B of the difference amplifier shown in Fig. 2.12, is given by Eq. (2.19). From this equation, if the relation $R_1/R_2 = R_3/R_4$ is observed,

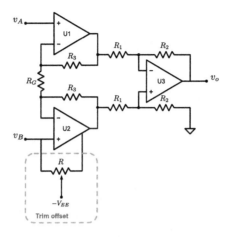

FIGURE 2.13

Classic instrumentation amplifier (pre-amplification).

the amplification gain can be set by choosing the ratio of resistors R_2 and R_1:

$$v_o = \frac{R_2(1 + R_1/R_2)}{R_1(1 + R_3/R_4)}v_A - \frac{R_2}{R_1}v_B. \tag{2.19}$$

One disadvantage of the circuit in Fig. 2.12 is that since voltage followers have unity gain, all common-mode rejection is left for the differential amplifier. This can only be accomplished by very precise resistor matching (Horowitz and Hill, 1989). Fig. 2.13 presents a circuit with a clever solution for this problem, which is known as the classic *instrumentation amplifier*. Op-amps U1 and U2 are used to raise the signal amplitude and also to reduce the representation of any potential that both sensors may have in common, relative to the ground of the power supply (Macfarlane et al., 2010a). Thus the output represents a signal with substantial reduction in common-mode signal when compared to the circuit in Fig. 2.12. If $R_1 = R_2$, then the input–output relation in Fig. 2.19 is given by Eq. (2.20):

$$v_o = \left(1 + \frac{2R_3}{R_G}\right)(v_B - v_A). \tag{2.20}$$

Note that the gain $G = 1 + 2R_3/R_G$ in Eq. (2.20) multiplies both AC and DC components of $(v_B - v_A)$. As a consequence, if DC component is large, it may saturate op-amp U3. For this reason, in ECG applications, it is convenient to place a potentiometer at input v_B to trim offsets produced by half-cell potentials. An integrated circuit with the instrumentation amplifier shown in Fig. 2.13 is available from several manufacturers. Typical examples are the low-power, general purpose INA128/129, and the high-precision AD624.

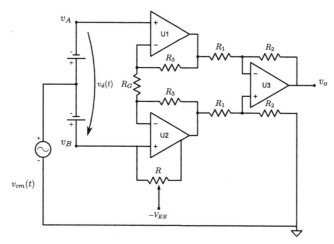

FIGURE 2.14

Common-mode rejection ratio.

2.4.2 COMMON MODE REJECTION

In an electric circuit it is common that the absolute potential of input v_A change by the same amount of input v_B, as illustrated in Fig. 2.14. The instrumentation amplifier tries to amplify $v_B - v_A$, but this difference fluctuates over time with respect to ground with an unknown voltage $v_{cm}(t)$. That is exactly what happens when we wish to measure lead potential, but body potential fluctuates randomly with respect to ground, because the body acts like an antenna capturing surrounding electromagnetic radiation.

In the ideal case, the output of a difference amplifier would be a function of the difference $v_B + \delta v_{cm} - (v_A + \delta v_{cm}) = v_A - v_B$, only. However, the output of a real amplifier is given by Eq. (2.21):

$$v_o = G_D(v_B - v_A) + G_C \frac{(v_B + v_A)}{2}, \tag{2.21}$$

where G_D is the differential gain, and G_C is the common-mode gain. It means that a small fraction of the average potential between the inputs is unintentionally added to the output. This happens because resistor matching is difficult to be exactly fulfilled. For instance, in differential amplifier (Fig. 2.12), if $R_1/R_2 - R_3/R_4 = 0$ is not exactly zero, then the term $G_C(v_B + v_A)/2$ is added to Eq. (2.20).

Common Mode Rejection Ratio (CMRR) is a figure of merit created to indicate how good an amplifier can reject a portion of the average inputs that is unintentionally added to the output. It is defined in Eq. (2.22) as the ratio between differential gain G_D and the common-mode gain G_C. Alternatively, this quantity is also expressed in decibels, such as in Eq. (2.23). In this case, it is referred to simply as Common Mode Rejection (CMR). Further information about CMRR in operational amplifiers can be

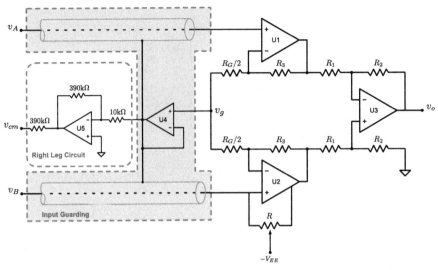

FIGURE 2.15

Input cables driven shield.

found in Webster (1998).

$$\text{CMRR} = \frac{G_D}{G_C}, \tag{2.22}$$

$$\text{CMR} = 20\log(\text{CMRR}). \tag{2.23}$$

2.4.3 INPUT GUARDING AND RIGHT LEG CIRCUIT

Electromagnetic interference is present everywhere, specially in hospital environments. Since human body is electrically conductive, an electrical current i_c may be induced by electromagnetic waves that reaches the body. If body resistance to ground is R_0, then a common-mode interference potential $v_{cm} = i_c R_0$ is developed (Webster, 1998). This electric potential is present in both leads of a differential electrode. Part of this interference is rejected by instrumentation amplifier, but a further improvement is possible by guarding the input cables.

If gain resistor R_G of Fig. 2.13, is split into two resistors of equal resistances $R_G/2$, we form a voltage divider with the average potential between v_A and v_B at node v_g. The input guarding can be done by putting a buffer to follow this voltage at the metallic shield of a coaxial cable. This situation is illustrated in Fig. 2.15 and is used to reduce the effects of cable capacitance and leakage.

Another improvement is achieved by what is known in the literature as *right leg circuit*. The right leg circuit actively removes the common-mode voltage v_{cm} from the body using the principle of negative feedback (Fig. 2.15). The key idea is to invert and amplify the output of the buffer and feed it back again into the body, with an inverter op-amp. The right leg lead, by standard convention, is used as the ground or the circuit reference (Webster, 1998).

2.4.4 **ANALOG FILTERING**

Even using cable guarding and feeding the average of measured potential back to the body, with the right leg circuit, some noise is inevitably picked up during the pre-amplification stage (Fig. 2.11). It means that the amplifier will multiply both signal and noise. This is one of the reasons to divide amplification in two stages. If only one amplification stage were used, noise would be amplified as much as the signal, which would result in a poor SNR.

Clifford et al. (2006) recommend to use a gain of 25x in the first stage for an adequate SNR. The amplification circuit of the second stage is a simple non-inverting op-amp. This second amplification stage further increases the SNR of the signal and can be set to boost the signal voltage to an amplitude adequate to the dynamic range of A/D converter. The high-pass filter is preferably placed between the first and second amplification stages.

The High-Pass Filter

ECG signals are commonly contaminated by half-cell potential differences at the skin-electrode interface (DC offset), changing also electrode impedance (Macfarlane et al., 2010a). This can be caused by electrode motion as a result of patient's breathing, movement, muscle tremor, temperature or skin moisture changes and gel leakage (Macfarlane et al., 2010a).

The changes at the skin-electrode interface result in ECG potential fluctuations that have more content in low frequency spectrum than those presented in a "standard" ECG. In time domain, it is observed as slow baseline wandering. This phenomenon can be attenuated by a high-pass filter placed after the first amplifier stage. There are several ways to implement high and low-pass analog filters and 2nd amplification stage. A simple way consists of using passive analog first-order RC filters in series with a non-inverting op-amp (Fig. 2.16). One can choose cutoff frequencies (fc_a and fc_b) by tunning resistors and capacitors such that $fc_a = 1/(2\pi R_a C_a)$ and $fc_b = 1/(2\pi R_b C_b)$, and defining the gain of non-inverting amplifier as $G = 1 + R_2/R_1$. One of the drawbacks of this design is the poor roll-off of first-order filters (-20 dB/decade). If Nyquist criterion were strictly met, that is, sampling frequency is twice the cutoff frequency of low pass filter, some aliasing may happen.

The transient response of a filter is also as important as its frequency response, since the accuracy of the QRS complex must be preserved for correct patient diagnosis. For example, the more the filter response is far from an ideal filter, the less sharp R-peak in QRS complex may be. Additionally, if filter frequency response is far from an ideal filter, oscillations near cut-off frequency (ringing) may be present. The linear phase active Bessel filter is a suitable filter for ECG applications because its characteristic frequency response is free from ringing and have excellent linear

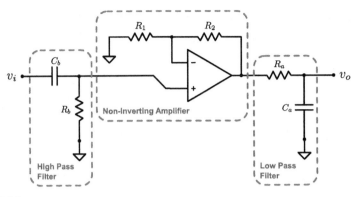

FIGURE 2.16

Analog filtering with 1st order passive filters and 2nd amplification stage with non-inverting amplifier.

phase (Northrop, 2004). Although these characteristics contribute to preserve ECG morphology, this filter suffers from slow steepness at the transition band. To overcome this limitation, higher orders can be used and for each increment in filter-order, the roll-off steepness is increased by -20 dB (Northrop, 2004). Clifford et al. (2006) recommend an eight-order Bessel high-pass filter in this stage.

The desired cutoff frequency of high-pass filter depends on the particular application involved. The suggested frequency-filtering band for the ECG signal, without introducing disturbance, falls between 0.05 Hz and 250 Hz (Kligfield et al., 2007).

The Low-Pass Filter

Skeletal muscles electrical activity can add undesirable components to ECGs, especially during exercise-stress testing, or under 24/7 continuous monitoring (Holter monitoring). Most of the spectral components of these sources are above the ECG frequency spectrum content and their effects could be attenuated by a low-pass filter (Kligfield et al., 2007; Clifford et al., 2006; Macfarlane et al., 2010a). If adult patients are under monitoring, the cutoff of this low-pass filter could be 150 Hz (Kligfield et al., 2007). The analog-type filter can also be the Bessel filter since this filter minimizes phase distortion (Northrop, 2004; Clifford et al., 2006). Another advantage of the low-pass filter implementation is reducing the influence of the aliasing effect that crops up at the digital ECG processing derived from the analog amplifier (Macfarlane et al., 2010a).

2.4.5 ELECTRIC HAZARD PROTECTION

ECG equipment exposes patients to electrical hazards because the electrode is in direct contact with the patient's skin and electrode-skin interface resistance are significantly reduced by electrode gel. It may create a low impedance path between power line and patient, increasing the risk of electrical shock in case of ECG electrical circuitry failure. A second hazard mechanism is when the operator simultaneously touches metallic parts of equipment and patient. Since body impedance is finite, any leaked current may flow from the rack of equipment to the patient, passing through the arms of operator and return to ground via patient body.

These problems can be addressed with the use of isolation amplifiers. These devices can be used to break up ground loops, eliminate source ground connections, provide patient isolation protection from power line and protect ECG equipment from high-input voltages produced by defibrillators or other electrosurgical equipment. The most common alternatives to promote isolation consist of applying transformers, capacitors, or optocouplers. The basic idea is to separate the front-end circuitry (that is, the part of the ECG system which is in direct contact with patient) from back-end circuitry (which is connected to power line and/or with a computer or monitor). Then the supply of front-end circuitry can be done by rechargeable batteries while the back-end circuit is supplied by power line.

A third failure mechanism happens when ECG equipment is in simultaneous use with other medical equipment. Defibrillators and other electrosurgical equipment applies a variable current that may induce large transient voltage at ECG electrode, which can damage amplifiers and conditioner circuit. Current to patient can be limited by placing 1 W resistors in lead cable. It is also recommended to connect voltage limiting devices between each measuring electrode and electric ground. These devices can be: (i) silicon diodes in parallel, (ii) Zener diodes, connected back-to-back or (iii) gas discharge tubes. They all appear as an open circuit when voltage exceeds a specified limit (Bronzino, 1999).

Safety standards for medical equipment may differ from country to country. The current organization that prepares and publishes International Standards for all electrical, electronic, and related technologies is the International Electrotechnical Commission (IEC), located in Geneva, Switzerland. It publishes standards for basic safety and essential performance of regular electrocardiographs (IEC 60601-2-25:2011, 2011), electrocardiographic monitoring equipment (IEC 60601-2-27:2011, 2011), and ambulatory electrocardiographic systems (IEC 60601-2-47:2012, 2012). The reader is referred to these documents for detailed requirements of safe ECG systems design.

2.5 CURRENT TECHNOLOGICAL TRENDS FOR ECG INSTRUMENTATION

Advances in biopotential sensors, ECG microelectronics chips, and signal processing techniques have made important contributions for the advances of ECG monitoring systems and diagnosis. Efforts have been made towards reduced size, power, and overall cost allowing great mobility of these systems within hospitals and by healthcare professionals at patients' home. These reduced hardwares, when connected into PCs or gadgets by USB, became a resting ECG system. Recent wireless ECG acquisition systems have shown that ECG signal integrity is preserved without the need of the patient being connected by cable to the system (Fornell, 2015).

Holter monitoring will be performed by inexpensive wearable (or even disposable) connected to smartphone-based ECG monitors. These devices, combined with app-based ECG monitors that are in immediate view of the patient, will spread the basic concepts of ECG monitoring, representing an important medical paradigm shift, higher degree of patient engagement (Fornell, 2015). This technology will impact in patient triage and early diagnosis.

Several current technologies now include embedded automatic algorithms that are based on patient-specific clinical information, ECG morphology, and ECG complexes segmentation to auxiliary medical decision on diagnosis and treatment, which can be exchanged by standard-based data exchange communication systems eliminating the need for paper-based ECG systems. By far, a large amount of ECG data is being generated and stored in datacenters, which in the future could help in decision-making of optimized signal processing techniques to improve automatic diagnosis, information security, and standardize ECG formats for interoperability in health IT systems (Fornell, 2015).

2.6 CONCLUSIONS

This chapter presented an overview of how an ECG signal acquisition system is organized. First, the phenomena related to electrode-skin interface and the associated biochemical effects involved, such as half-cell potential and electrical modeling of interface, was shown. Then the different common lead systems were accessed for ECG-related diagnosis, monitoring, and exercise. The main components of ECG signal conditioner were also presented with some electronic design guidelines for signal amplification, filtering, and isolation. The chapter ended with an overview of new trends for ECG instrumentation.

REFERENCES

Bronzino, J.D., 1999. Biomedical Engineering Handbook, vol. 2. CRC Press.

Clifford, G.D., Azuaje, F., McSharry, P., et al., 2006. Advanced Methods and Tools for ECG Data Analysis. Artech House, Norwood, MA.

Edenbrandt, L., Pahlm, O., 1988. Vectorcardiogram synthesized from a 12 lead ECG: superiority of the inverse Dower matrix. Journal of Electrocardiology 21 (4), 361–367.

Einthoven, W., Fahr, G., de Waart, A., 1950. On the direction and manifest size of the variations of potential in the human heart and on the influence of the position of the heart on the form of the electrocardiogram. American Heart Journal 40 (3), 163–211.

Fornell, D., 2015. Advances in ECG technology September 16.

Frank, E., 1956. An accurate, clinically practical system for spatial vectorcardiography. Circulation 13 (5), 737–749.

Gari, D.C., Francisco, A., Patrick, E., 2006. Advanced Methods and Tools for ECG Data Analysis. Artech House, Inc.

Goldberger, E., 1942. A simple, indifferent, electrocardiographic electrode of zero potential and a technique of obtaining augmented, unipolar, extremity leads. American Heart Journal 23 (4), 483–492.

Horowitz, P., Hill, W., 1989. The Art of Electronics. Cambridge Univ. Press.

IEC 60601-2-25:2011, 2011. Medical electrical equipment – Part 2-25: particular requirements for the basic safety and essential performance of electrocardiographs. Standard, International Electrotechnical Commission. Geneva, Switzerland.

IEC 60601-2-27:2011, 2011. Medical electrical equipment – Part 2-27: particular requirements for the basic safety and essential performance of electrocardiographic monitoring equipment. Standard, International Electrotechnical Commission. Geneva, Switzerland.

IEC 60601-2-47:2012, 2012. Medical electrical equipment – Part 2-47: particular requirements for the basic safety and essential performance of ambulatory electrocardiographic systems. Standard, International Electrotechnical Commission. Geneva, Switzerland.

Kligfield, P., Gettes, L.S., Bailey, J.J., Childers, R., Deal, B.J., Hancock, E.W., van Herpen, G., Kors, J.A., Macfarlane, P., Mirvis, D.M., Pahlm, O., Rautaharju, P., Wagner, G.S., 2007. Recommendations for the standardization and interpretation of the electrocardiogram: Part I: the electrocardiogram and its technology a scientific statement from the American Heart Association Electrocardiography and Arrhythmias Committee, Council on Clinical Cardiology; the American College of Cardiology Foundation; and the Heart Rhythm Society endorsed by the International Society for Computerized Electrocardiology. Journal of the American College of Cardiology 49 (10), 1109–1127.

Macfarlane, P.W., van Oosterom, A., Pahlm, O., Kligfield, P., Janse, M., Camm, J., 2010a. Comprehensive Electrocardiology, vol. 1. Springer Science & Business Media.

Macfarlane, P.W., van Oosterom, A., Pahlm, O., Kligfield, P., Janse, M., Camm, J., 2010b. Comprehensive Electrocardiology, vol. 3. Springer Science & Business Media.

Northrop, R.B., 2004. Analysis and Application of Analog Electronic Circuits to Biomedical Instrumentation. CRC Press LLC.

Ramanathan, C., Ghanem, R.N., Jia, P., Ryu, K., Rudy, Y., 2004. Noninvasive electrocardiographic imaging for cardiac electrophysiology and arrhythmia. Nature Medicine 10 (4), 422.

Taccardi, B., Punske, B.B., Lux, R.L., Macleod, R.S., Ershler, P.R., Dustman, T.J., Vyhmeister, Y., 1998. Useful lessons from body surface mapping. Journal of Cardiovascular Electrophysiology 9 (7), 773–786.

Uijen, G.J.H., van Oosterom, A., van Dam, R.T., 1988. The relationship between the 12-lead standard ECG and the XYZ vector leads. In: 14th Int. Congr. Electrocardiology. IEEE, pp. 301–307.

van der Graaf, A.W.M., Bhagirath, P., Ramanna, H., van Driel, V.J., de Hooge, J., de Groot, N.M., Götte, M.J.W., 2014. Noninvasive imaging of cardiac excitation: current status and future perspective. Annals of Noninvasive Electrocardiology 19 (2), 105–113.

Van Oosterom, A., Oostendorp, T., 2004. ECGSIM: an interactive tool for studying the genesis of QRST waveforms. Heart 90 (2), 165–168.

Webster, J.G., 1998. The Measurement, Instrumentation and Sensors Handbook. CRC Press.

Webster, J.G., 2006a. Encyclopedia of Medical Devices and Instrumentation, vol. 3. John Wiley & Sons, Inc., Publication.

Webster, J.G., 2006b. Encyclopedia of Medical Devices and Instrumentation, vol. 1. John Wiley & Sons, Inc., Publication.

Willems, J., 1984. Common standards for quantitative electrocardiography. In: 4th Progress Report. IEEE, pp. 199–200.

Techniques for Noise Suppression for ECG Signal Processing

João Paulo do Vale Madeiro*, Paulo César Cortez†,
José Maria da Silva Monteiro Filho‡, Priscila Rocha Ferreira Rodrigues‡
*Institute for Engineering and Sustainable Development (IEDS), University for the International
Integration of the Afro-Brazilian Lusophony – UNILAB, Redenção, Ceará, Brazil
†Department of Teleinformatics Engineering, Federal University of Ceara, Fortaleza, Ceará, Brazil
‡Department of Computing Science, Federal University of Ceara, Fortaleza, Ceará, Brazil

3.1 SPECTRAL CONTENT OF THE ECG SIGNAL VERSUS NOISE SPECTRA

The ECG signal contains, in addition to the clinical and cardiac physiological content related to QRS complex, P- and T-waves, noise content related to 60-Hz noise from power line interference, electromyography (EMG) signal from skeletal muscle activity, motion artifact from the electrode and skin interface, and other interference sources from electrosurgery equipment in the clinic or operating room (Tompkins, 1993). Despite the inherent and typical morphological variations within the ECG characteristic waves, from beat to beat in a single patient record, and more evidently from patient to patient, QRS complex, P-wave and T-wave are analyzed as distinctive features, and automatic identification of these features is, to some extent, a tactable issue. However, removing and even quantifying the noise information in the ECG is not straightforward and not always an achievable task. This is partially due to the fact that there are so many different types of interference and artifacts that can occur simultaneously, and also because these noises are often transient, and unpredictable in terms of their onset and duration (Gari et al., 2006).

In summary, ECG signal noise can be classified as (Friesen et al., 1990; Gari et al., 2006):

- Power line interference: 60 Hz mains noise;
- Electrode pop or contact noise: loss of contact between the electrode and the skin manifesting as sharp changes;
- Patient-electrode motion artifacts: movement of the electrodes away from the contact area on the skin, leading to variations in the impedance between the electrode and skin, causing potential variations in the ECG and usually manifesting themselves as rapid baseline jumps or complete saturation for up to 0.5 s;

Developments and Applications for ECG Signal Processing. https://doi.org/10.1016/B978-0-12-814035-2.00009-8

- EMG noise: electrical activity due to muscle contractions with spectral content between 0 and 10,000 Hz, with an average amplitude of 10% of the full scale deflection;
- Baseline drift: usually from respiration activity with an amplitude of around 15% of full scale deflection at frequencies between 0.15 and 0.3 Hz.

Aiming to discriminate the power-spectra content individually related to QRS complex, P- and T-waves from muscle noise and motion artifacts, Thakor et al. (1984) obtained: (1) normal ECGs from resting healthy subjects; (2) ECGs with muscle noise caused by flexing the arm and the chest muscles; (3) motion artifacts from subjects jogging on a treadmill, and (4) abnormal ECGs. After acquiring 150 beats in each category from six subjects, the authors obtained a detailed graphic analysis showing the relative power spectra of 150 complete noise-free ECG cycles, the QRS complexes, and the P- and T-waves, as can be observed in Thakor et al. (1984).

According to their study, as observed in Thakor et al. (1984), clinical ECG spectral content extends from around 2.5 Hz to around 40 Hz. Muscle noise superimposes frequency content related to both QRS complex and P/T-waves, whereas motion artifact superimposes, basically, frequency content related to P/T-waves.

Concerning QRS complex, which is the most emphasized ECG characteristic wave and being that its detection is the first step in ECG automatic analysis, it has been demonstrated that Murthy et al. (1978):

- In a typical QRS of normal duration, virtually all of the power is contained in frequencies below 30 Hz with peak power occurring in the range of 4 to 12 Hz;
- Premature ventricular contractions (PVC's) typically contain less high-frequency power than normal beats, with virtually all the power contained in frequencies below 12 Hz with peak power at about 4 Hz;
- Notches in the QRS are accompanied by a definite broadening of the power-spectral density with significant power present in the range 30–100 Hz, although the additional power is still much smaller than that contained in the band below 30 Hz.

3.2 THE BASIS OF ECG FILTERING

Most waveforms consist of signal plus noise mixed together. Signal and noise are relative terms, relative to the task at hand: the signal is that portion of the waveform of interest, whereas the noise is everything else (Semmlow and Griffel, 2014). In general, the goal of ECG filtering is to separate out the pure ECG signal from noise or to detect features of an ECG signal buried in noise.

As a basic concept, the relative amount of signal and noise present in a waveform is usually quantified by the signal-to-noise ratio (SNR), which is simply the ratio of signal to noise, both measured in RMS (root-mean-squared) amplitude and often expressed in "db" as given by Eq. (3.1).

$$SNR = 20 \log \frac{Signal}{Noise}.$$
(3.1)

As already explained in the previous chapter, analog-to-digital conversion (ADC) (or time sampling) transforms a continuous analog signal into a discrete time signal, which can be thought of as an array in computer memory: $x(n) = [x_1, x_2, x_3, ..., x_N]$. The array position indicates a relative position in time, so that converting back to relative time is then is achieved by multiplying the index number n by the sampling interval, $T_s : x(t) = x(nT_s)$.

According to sampling theory, a sinusoid can be uniquely reconstructed providing it has been sampled by at least two equally spaced points over a cycle. Since Fourier series analysis implies that any signal can be represented as a series of sine waves, by extension, a signal can be uniquely reconstructed, providing the sampling frequency is twice that of the highest frequency in the signal (Semmlow and Griffel, 2014). On the other hand, any signal that contains frequency components greater than a half of the sampling frequency cannot be reconstructed, and its digital representation is in error, which is termed aliasing, and no amount of digital signal processing can correct this error.

Aliasing must be avoided either by the use of sampling rates well above the bandwidth of the original analog signal or by filtering the analog signal before analog-to-digital conversion. Since very high sampling rates implies memory costs, and also in terms of ADC, the filtering approach is generally preferable. Therefore all frequencies in the sampled waveform greater than one half the sampling frequency must be sufficiently attenuated (less than quantization noise level or other acceptable noise level) (Semmlow and Griffel, 2014).

A given signal in the time domain can mathematically be represented in terms of its magnitude and phase responses in the frequency domain. As illustrative examples, Fig. 3.1 presents an excerpt of a real 5-s ECG signal collected through the shield Olimex board and model SHIELD EKG-EMG.

The shield Olimex board and model SHIELD-EKG-EMG is an open source hardware, which allows Arduino-like boards to capture electrocardiography signals. The shield converts the analog differential signal (ECG biopotentials) into a single stream of data as output. A third-order "Besselworth" analog filter's cutoff frequency is set to $f_c = 40$ Hz. The output signal is discretized via a dedicated ADC embedded in the Arduino board. Default values for discretizing parameters are: 10-bit ADC with 256 Hz sampling rate. The electrodes used for signal acquisition consist of three leads, which should be placed on the wrists and right ankle of the patient (Villarrubia et al., 2014).

Figs. 3.2 and 3.3 present the frequency response, including both magnitude and phase responses, for the referred excerpt of ECG signal (Fig. 3.1), which are computed using Fast Fourier Transform Algorithm (Welch, 1967). Then Fig. 3.4 illustrates the power spectrum of the same signal, which is commonly defined as the Fourier Transform of the autocorrelation function.

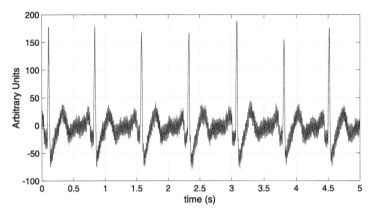

FIGURE 3.1

An excerpt of a 5-s ECG signal collected through the shield Olimex board and model
SHIELD EKG-EMG.

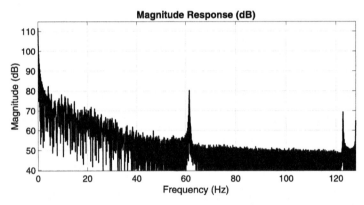

FIGURE 3.2

Magnitude response for an excerpt of a 5-s ECG signal collected through the shield Olimex
board and model SHIELD EKG-EMG.

The filtering of a time series involves the transformation of a discrete one-dimensional signal $x[n]$, consisting of N points, such that $x[n] = (x_1, x_2, x_3, ..., x_N)^T$, into a new representation, $y[n] = (y_1, y_2, y_3, ..., y_N)^T$. If $x[n]$ is a column vector that represents a channel of ECG, we can generalize this representation so that the input data $\mathbf{X}$, composed by M channels of ECG, and the corresponding transformed representation $\mathbf{Y}$ are given by Gari et al. (2006)

$$X = \begin{bmatrix} x_{11} & x_{12} & \cdots & x_{1M} \\ x_{21} & x_{22} & \cdots & x_{2M} \\ \cdots & \cdots & \cdots & \cdots \\ x_{N1} & x_{N2} & \cdots & x_{NM} \end{bmatrix}, Y = \begin{bmatrix} y_{11} & y_{12} & \cdots & y_{1M} \\ y_{21} & y_{22} & \cdots & y_{2M} \\ \cdots & \cdots & \cdots & \cdots \\ y_{N1} & yN2 & \cdots & y_{NM} \end{bmatrix}, \mathbf{Y}^T = \mathbf{W}\mathbf{X}^T, \quad (3.2)$$

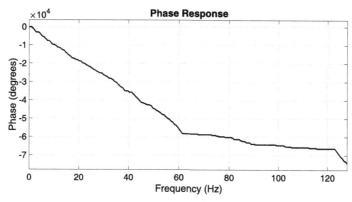

FIGURE 3.3

Phase response for an excerpt of a 5-s ECG signal collected through the shield Olimex board and model SHIELD EKG-EMG.

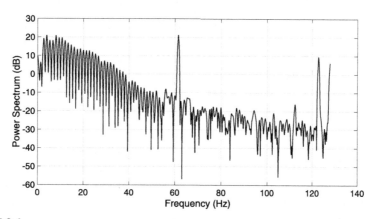

FIGURE 3.4

Power spectrum for an excerpt of a 5-s ECG signal collected through the shield Olimex board and model SHIELD EKG-EMG.

where an $M \times M$ transformation matrix $\mathbf{W}$ is applied to $\mathbf{X}$ to create the transformed matrix $\mathbf{Y}$.

The main goal of a transformation is to project data into another space, which serves to highlight patterns in data. Highlighting data patterns has to do with filtering data, that is, discarding noise, or uninteresting parts of the signal, which are superimposing the clinical and focused information. As we are discarding the dimensions that correspond to the noise, we are proceeding with a dimensionality reduction (Gari et al., 2006).

Now we will detail a discussion concerning accuracy of ECG preprocessing filters in the context of guaranteeing an efficient ECG interpretation and diagnosis, covering three specific ECG preprocessing applications: baseline wander suppression, line frequency rejection, and muscle artifact reduction.

3.2.1 BASELINE WANDER SUPPRESSION

Baseline wander noise may be associated with respiratory activity and perspiration, and increased or sudden body movement. Removal of this kind of interference is required in order to minimize changes in beat morphology, which do not have cardiac origin (Sörnmo and Laguna, 2005). For example, subtle changes in the ST–T segment are analyzed for diagnosis of ischemia. However, these changes may be confused with baseline wandering during the course of a stress test (Singh, 1987).

The frequency content of baseline wander is commonly in the range of 0 to 0.5 Hz. Some events considerably increase the frequency content of baseline wander, such as increased movement of the body during the latter stages of a stress test or abrupt movements. There are two important recommendations for the technical requirements of lower-frequency cutoff in ECG instruments. The first is the 1975 American Heart Association (AHA) recommendations (Pipberger et al., 1975). Based on an analog single-pole filter with insignificant distortion of the ST segment and QT interval, the committee recommended a lower-frequency cutoff (3-dB down) (Berson and Pipberger, 1966). The second is the 1990 AHA recommendations by an Ad Hoc Writing Group Committee (Bailey et al., 1990). Considering that the longest RR interval corresponds to the lowest frequency components of the ECG, this document recommended a lower-frequency cutoff (3-dB down) at 0.67 Hz, which corresponds to a heart rate of 40 beats per minute. Also, it requires less than 0.5-dB ripple over the range of 1 to 30 Hz.

According to Sörnmo and Laguna (2005), the two major techniques employed for the removal of baseline wander from the ECG are linear filtering and polynomial fitting, being the first family of techniques divided into filtering, based on time-invariant or time-variant structures.

Considering the design of linear, time-invariant high-pass filters, the cutoff frequency should obviously be chosen, so that clinical information remains undistorted while, as much as possible, the baseline wander is suppressed. The lowest frequency component of the ECG may be identified as the one associated to a heart rate of 40 beats per minute, which is a candidate for low cutoff frequency $F_c = 0.67$ Hz. Regarding regular fluctuations around the heart rate of 40 beats per minute, it is necessary to choose a slightly lower cutoff frequency of approximately $F_c = 0.5$ Hz.

Beyond the definition of cutoff frequency, the phase response is another decision-making element, given that linear phase filtering is highly desirable in order to prevent phase distortion from changing wave features, such as morphology, duration, and amplitude of the QRS complex, P- and T-waves, or the end point of the T-wave. In this context, it is well-known that Finite Impulse Response (FIR) filters can provide an exact linear phase response, and Infinite Impulse Response (IIR) filters introduce distortions due to its nonlinear phase response. However, as it will be described and illustrated in the following subsection, for a linear phase FIR filter to work with satisfactory accuracy, a very long impulse response is necessary.

3.2.1.1 Linear and Time-Invariant Filtering: FIR and IIR Filters

A basic and direct approach to design a FIR high-pass filter is to parameterize an ideal filter as a starting point, as defined by Eq. (3.3):

$$H(e^{j\omega}) = \begin{cases} 0, 0 \le |\omega| \le \omega_c, \\ 1, \omega_c < |\omega| < \pi. \end{cases} \tag{3.3}$$

The corresponding impulse response has an infinite length (Oppenheim, 1999), as defined by Eq. (3.4):

$$\begin{aligned} h[n] &= \frac{1}{2\pi} \int_{\omega_c}^{\pi} 1 \cdot e^{j\omega n} d\omega + \frac{1}{2\pi} \int_{-\pi}^{-\omega_c} 1 \cdot e^{j\omega n} d\omega \\ &= \begin{cases} 1 - \frac{\omega_c}{\pi}, n = 0, \\ \frac{-\sin(\omega_c n)}{\pi n}, n = \pm 1, \pm 2, \dots. \end{cases} \end{aligned} \tag{3.4}$$

Truncation of $h[n]$ can be done by multiplying it by a rectangular window function, named $w_r[n]$, as demonstrated by Eq. (3.5):

$$w_r[n] = \begin{cases} 1, |n| = 0, \dots, L, \\ 0, |n| > L, \end{cases} \tag{3.5}$$

or by another window function, as Hamming window, given by $w_h[n]$ (see Eq. (3.6)).

$$w_h[n] = 0.54 - 0.46 \cos(\frac{2\pi k}{N-1}), k = 0, 1, \dots, N-1. \tag{3.6}$$

Concerning the evaluation for applying FIR high-pass filter to suppress baseline wander, we may use an alternative approach for generating synthetic ECG signals with arbitrary morphologies (any lead configuration), ECGSYN, which is available on the PhysioNet website (Goldberger et al., 2000). ECGSYN is based upon time-varying differential equations and is continuous, with convincing beat-to-beat variations in each waveform morphology and also in interbeat intervals (Gari et al., 2006; McSharry et al., 2003). An excerpt of a 5-s synthetic ECG signal, 360-Hz sampling frequency, reflecting the electrical activity in the heart during eight beats, is presented in Fig. 3.5. Regarding QRS complex morphology, we identify a typical "Rs" morphology. For a set of typical QRS morphologies, see the master thesis developed by Lugovaya (2005), also available from the PhysioNet website.

In Figs. 3.6(A), 3.6(B), 3.6(C), and 3.6(D), we can observe the magnitude frequency response of four different high-pass FIR filters using a Hamming window with variable length: 50, 500, 1000, and 1500 coefficients.

Concerning the evaluation of a high-pass FIR filter for filtering a simulated baseline wander noise, in Fig. 3.7, we observe a very poor noise suppression of a 0.6-Hz frequency sinusoidal noise, considering cutoff frequency of 0.5 Hz and a Hamming window with 1500 coefficients.

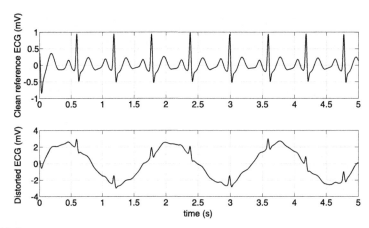

FIGURE 3.5

An excerpt of a 5-s synthetic ECG signal, reflecting the electrical activity in the heart during eight beats (upper panel), and the contaminated version of the same excerpt by a sinusoidal noise, having 0.6-Hz frequency and 2.5-mV amplitude.

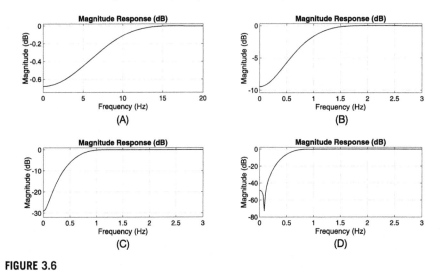

FIGURE 3.6

Magnitude frequency response for high-pass FIR filters, using a Hamming window, with variable length: $50 \leq N \leq 1500$. (A) Magnitude response for $N = 50$ coefficients. (B) Magnitude response for $N = 500$ coefficients. (C) Magnitude response for $N = 1000$ coefficients. (D) Magnitude response for $N = 1500$ coefficients.

A number of techniques may be considered for solving the issues related to the poor performance and efficiency of FIR-filters for suppressing baseline wander, such as forward–backward IIR filtering and insertion of zeros into an FIR filter.

The use of forward–backward filtering remedies the disadvantage of nonlinear phase response of IIR filters, since the overall result is filtering with a zero-phase

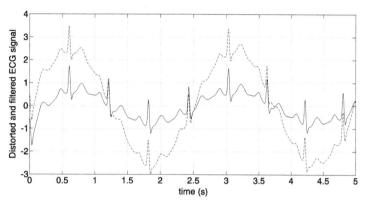

FIGURE 3.7

An excerpt of a 5-s synthetic ECG signal with a 0.4-Hz frequency sinusoidal noise (dashed line) and its filtered version (solid line), considering a high-pass FIR filter with a cutoff frequency of 0.5 Hz, and a Hamming window with 1500 coefficients.

transfer function. Due to the freedom of positioning its poles, an IIR filter achieves magnitude frequency-response requirements more easily with a much lower filter order. According to Sörnmo and Laguna, Oppenheim (2005, 1999), a forward–backward filtering involves three steps. The first stage refers to the processing of the input signal, named $x[n]$, with an IIR filter impulse response $h[n]$. Then the corresponding filter output is time reversed, that is, we create a reflected version over time axis. The third stage refers to another processing over the time-reflected filter output, also with $h[n]$. Finally, the second output (doubly filtered signal) is again time reversed to produce the output signal $y[n]$.

Representing the referred to stages through discrete-time mathematical expressions and convolution operators allow us to obtain (3.7), (3.8), and (3.9).

$$s_1[n] = h[n] * x[n], \tag{3.7}$$

$$s_2[n] = h[n] * s_1[-n], \tag{3.8}$$

$$y[n] = s_2[-n]. \tag{3.9}$$

Transforming the mathematical relations to the frequency domain, through discrete-time Fourier Transform, allows us to verify that the output $y[n]$ has the same phase response of the input signal $x[n]$ (see Eq. (3.10)).

$$Y(e^{j\omega}) = S_2^*(e^{j\omega}) = H^*(e^{j\omega})S_1(e^{j\omega}) = H^*(e^{j\omega})H(e^{j\omega})X(e^{j\omega})$$
$$= |H(e^{j\omega})|^2 X(e^{j\omega}). \tag{3.10}$$

Therefore although $h[n]$ itself has a nonlinear phase response, the input signal $x[n]$ is filtered by a transfer function, whose magnitude is squared $|H(e^{j\omega})|^2$, the filter order is doubled, and phase function is zero.

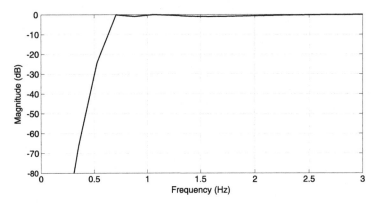

FIGURE 3.8

Magnitude frequency response for high-pass forward–backward, eighth-order type-II Chebyshev IIR filter with a cutoff frequency of 0.5 Hz.

The downsides of the approach related to forward–backward filtering are that the initial transients may be larger, making it inappropriate for many FIR filters that depend on phase response for proper operation (Semmlow and Griffel, 2014). Also, this approach is nonapplicable for strictly real-time ECG signal processing, but indeed is a scheme for off-line processing, since the requirement of causality is not attended when a time-reversed signal is processed. However, forward–backward IIR filtering can be implemented with a relatively short (time) delay as a result of processing consecutive and overlapping signal segments, which may be considered by real-time applications.

Fig. 3.8 presents the magnitude frequency response for a high-pass forward–backward eighth-order type-II Chebyshev IIR filter with a passband frequency of 0.6 Hz/360 Hz. We can observe a deep and fast transition from stopband to passband frequency, with a more efficient attenuation of frequencies close to zero, and also a stable behavior for magnitude response at passband frequency range.

Concerning the referred to IIR filter, Fig. 3.9 presents an excerpt of a 5-s synthetic ECG signal with a 0.4-Hz frequency sinusoidal noise (dashed line) and its filtered version (solid line).

Another technique for improving filtering efficiency related to attenuation of baseline wandering is the insertion of zeros (or zero-padding) into a finite impulse response $h_0[n]$, which should be originally designed for a much lower sampling rate F_{s_0} (Sörnmo and Laguna, 2005).

Therefore the proper operating condition for inserting $D-1$ zeros in between every sample in $h_0[n]$ for obtaining a new $h[n]$ impulse response is given by expression (3.11).

$$h[n] - \begin{cases} h_0[n/D], |n| = 0, D, 2D, ..., \\ 0 \text{ otherwise.} \end{cases} \tag{3.11}$$

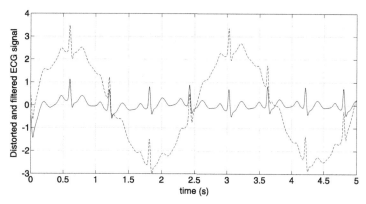

FIGURE 3.9

An excerpt of a 5-s synthetic ECG signal with a 0.4-Hz frequency sinusoidal noise (dashed line) and its filtered version (solid line), considering a high-pass, forward–backward, eight-order type-II Chebyshev IIR filter with a passband frequency of 0.6 Hz.

The effect of that operation is the D-fold repetition of the original transfer function $H_0(e^{j\omega})$, which may be checked by calculating the discrete-time Fourier transform of $h[n]$:

$$H(e^{j\omega}) = \sum_{n=-\infty}^{\infty} h[n]e^{-j\omega n} = \sum_{n=-\infty}^{\infty} h_0[n]e^{-j\omega n D} = H_0(e^{j\omega D}). \qquad (3.12)$$

The effect of zero-padding is illustrated by Fig. 3.10 and Figs. 3.11(A) and (B). The first figure illustrates the magnitude response function $H_0(e^{j\omega})$, originally related to a high-pass FIR filter with a cutoff frequency of 0.5 Hz, and a Hamming window with 500 coefficients for a sampling frequency of 60 Hz. The second figure illustrates the magnitude response function $H(e^{j\omega})$, resulting from setting $D = 5$ for zero-padding (Fig. 3.11(A) shows detailed behaviour for range 0–3 Hz, and Fig. 3.11(B) shows detailed behaviour for range 50–70 Hz).

As emphasized by Sörnmo and Laguna (2005), the insertion of zeros into a finite impulse response attenuates not only baseline wander, but also signal frequencies present at multiples of the original sampling, which is 60 Hz for the present illustration. As an illustrative example, Fig. 3.12 presents an excerpt of a 5-s synthetic ECG signal, with a 0.4-Hz, plus 60-Hz frequency sinusoidal noise, and its filtered version, considering a FIR filter $h[n]$ after inserting zeros into the original impulse response $h_0[n]$.

Taking as an evaluation parameter for filtering routines the normalized root mean square (RMS) error, between the reference clean ECG signal $y[n]$ and a corresponding filtering version $\hat{y}[n]$, given by the expression

$$\varepsilon_{RMS} = \frac{\sum_k [(y[k] - \hat{y}[k])^2]}{\sum_k [(y[k]^2)]}, \qquad (3.13)$$

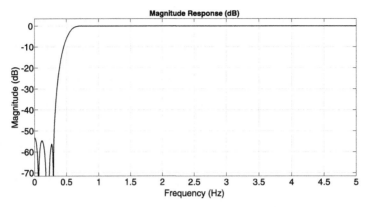

FIGURE 3.10

The original magnitude response function of a high-pass FIR filter $h_0[n]$ with a cutoff frequency of 0.5 Hz, and a Hamming window with 500 coefficients for a sampling frequency of 60 Hz.

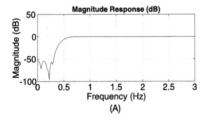
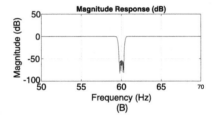

FIGURE 3.11

Magnitude response of a FIR filter $h[n]$ after inserting zeros into the original impulse response $h_0[n]$: (A) detailed behavior for range 0–3 Hz, and (B) detailed behavior for range 50–70 Hz.

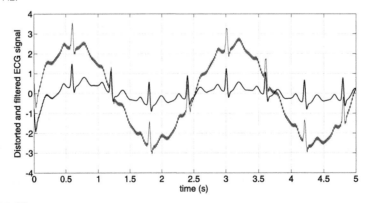

FIGURE 3.12

An excerpt of a 5-s synthetic ECG signal with a 0.4-Hz, plus 60-Hz frequency sinusoidal noise and its filtered version, considering a FIR filter $h[n]$ after inserting zeros into the original impulse response $h_0[n]$.

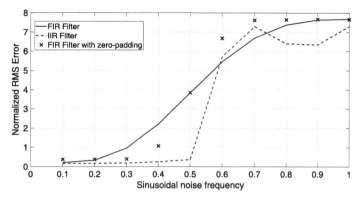

FIGURE 3.13

Performance comparison of three possible approaches for linear time-invariant filtering as sinusoidal noise frequency, ranging from 0.1 to 1 Hz, applying FIR filter (cutoff frequency of 0.5 Hz and Hamming window with 1500 coefficients), backward–forward IIR filtering (eighth-order type-II Chebyshev IIR filter, with a cutoff frequency of 0.5 Hz) and FIR filtering with zero-padding ($D = 6$), originally related to a high-pass FIR filter with a cutoff frequency of 0.5 Hz, and a Hamming window with 500 coefficients for a sampling frequency of 60 Hz.

we can compare the performance of filtering techniques based on FIR filters, forward–backward IIR filters, and FIR filtering with zero-padding as varying frequency of a sinusoidal noise. As we can see from Fig. 3.13, the approach based on FIR filtering with a cutoff frequency of 0.5 Hz, and a Hamming window with 1500 coefficients, presents the highest error values for sinusoidal noise frequency, ranging from 0.1 Hz to 0.5 Hz. The insertion of zero-padding ($D = 6$) allows obtaining a better performance for FIR filtering at the referred to range, considering an original high-pass FIR filter with a cutoff frequency of 0.5 Hz, and a Hamming window with 500 coefficients for a sampling frequency of 60 Hz. However, the more accurate results are obtained by forward–backward IIR filtering, considering an eighth-order type-II Chebyshev IIR filter with a cutoff frequency of 0.5 Hz.

3.2.1.2 Linear and Time-Variant Filtering

The major limitation in using time-invariant filtering for removing baseline wandering is its inefficient performance related to a fixed cutoff frequency, which is in turn based on the lowest possible heart rate. Since the human heart rate fluctuates, and also may achieve stable high levels, for example, during the latter stages of a stress test, it may be imperative to instantaneously adapt the cutoff frequency $f_c[n]$ according to the instantaneous "prevailing" heart rate, rather than to the lowest possible heart rate.

According to Sörnmo and Laguna (2005), the "prevailing heart rate" may be represented in terms of the instantaneous RR interval estimate, as heart rate is inversely proportional to RR interval. Therefore if two successive heartbeats (QRS complexes) occur at time instants R_i and R_{i+1}, the corresponding RR interval is computed as

$$RR[i + 1] = R[i + 1] - R[i]. \tag{3.14}$$

Considering a time-series composed by consecutive results for $RR[i]$, linear interpolation may be applied to synthesize a discrete-time series $r[n]$, for each time index n, as suggested by Sörnmo and Laguna (2005) (see Eq. (3.15)).

$$RR[n] = RR[i] + \frac{RR[i+1] - RR[i]}{R[i+1] - R[i]}(n - R[i]), n = R[i], ..., R[i+1]. \quad (3.15)$$

Considering the condition of inverse proportionality between time-varying cutoff frequency $f_c[n]$ as $f_c[n] = \frac{60}{RR[n]}$, a time-varying impulse response for a lowpass filter $h[k,n]$, aiming for the attenuation of baseline wandering, may be derived from the inverse discrete-time Fourier Transform of the frequency response of an ideal lowpass filter. The resultant expression is given as (Sörnmo and Laguna, 2005)

$$h[k,n] = \frac{1}{2\pi} \int_{-\omega_c[n]}^{\omega_c[n]} 1 \cdot e^{j\omega k} d\omega = \begin{cases} \frac{2\pi f_c[n]}{\pi}, k = 0; \\ \frac{\sin(2\pi f_c[n]k)}{\pi.k}, k \neq 0. \end{cases} \quad (3.16)$$

The index k denotes time within the impulse response, the index n denotes the time at which the filter should be applied, and $f_c[n]$ refers to the variable cutoff frequency.

3.2.1.3 Techniques Based on Polynomial Fitting

Techniques based on polynomial fitting are an alternative to approaches based on linear-filtering techniques, and consists of computing a polynomial to fit representative samples of the ECG, named "knots", where each "knot" is defined for each beat. Considering that filtered ECG signal will be computed by subtracting the derived polynomial from the original signal, the "knots" are selected from the isoelectric line, which may correspond to the PQ interval or the TP interval. For example, Lemay et al. (2005) apply baseline correction by means of a cubic spline interpolation anchored on QRS onset points (PQ interval). A common characteristic of these techniques is that their implementation requires the QRS complexes to be first detected and/or delineated such that "knots" may be accurately identified.

By simply connecting successive knots, which is equivalent to a first-order polynomial, the resulting baseline estimate is extremely poor and is not able to follow the fluctuations, and also its derivatives at the "knots" are discontinuous. By using higher-order polynomials, we achieve the estimation of more accurate baseline suppressions. A common approach involves applying third-order polynomial fitting to successive triplets of knots, which is referred to as cubic spline baseline estimation (De Boor, 1978; Sörnmo and Laguna, 2005; Lemay et al., 2005).

Fig. 3.14(A) presents an excerpt of a 6 s synthetic ECG signal, with a 0.5-Hz frequency sinusoidal noise (solid line) and its polynomial fitting for baseline wander removal (dashed line), based on spline cubic interpolation of QRS onsets. Fig. 3.14(B) presents the ECG signal resultant from the corresponding subtraction process. An algorithm for QRS detection and delineation, based on Wavelet and Hilbert Trans-

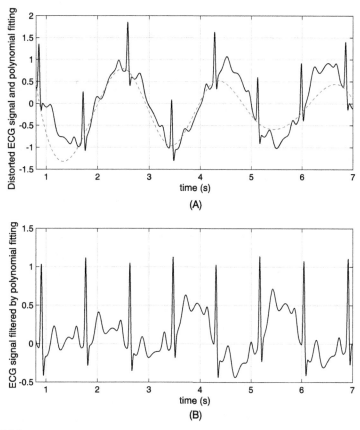

FIGURE 3.14

Filtering based on polynomial fitting: (A) an excerpt of a 5-s synthetic ECG signal with a 0.5-Hz frequency sinusoidal noise (solid line) and its polynomial fitting for baseline wander removal (dashed line), based on spline cubic interpolation of QRS onsets, and (B) the resultant subtracted ECG signal.

forms (Madeiro et al., 2012), is applied with the aim to achieve the knots required for polynomial fitting.

For comparison purposes, we proceed with spline cubic interpolation, also considering other fiducial points pertained to ECG isoelectric line: T-wave ends and P-wave onsets. An algorithm for detection and delineation of T-waves and P-waves, based on mathematical modeling, is applied to achieve these knots required for polynomial fitting.

Fig. 3.15(A) presents an excerpt of a 6-s synthetic ECG signal, with a 0.5-Hz frequency sinusoidal noise (solid line) and its polynomial fitting for baseline wander removal (dashed line), based on spline cubic interpolation of QRS onsets, T-wave

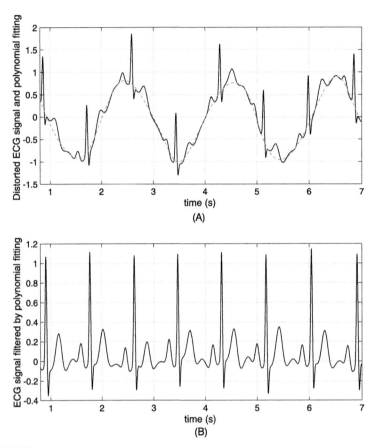

FIGURE 3.15

Filtering based on polynomial fitting: (A) an excerpt of a 5-s synthetic ECG signal, with a 0.5-Hz frequency sinusoidal noise (solid line) and its polynomial fitting for baseline wander removal (dashed line), based on spline cubic interpolation of QRS onsets, T-wave ends, and P-wave onsets and (B) the resultant subtracted ECG signal.

ends, and P-wave onsets. Fig. 3.15(B) presents the ECG signal resultant from the corresponding subtraction process.

Taking the normalized RMS error as an evaluation parameter, we can compare the performance of filtering techniques based on polynomial filtering, considering only QRS onsets as knots and considering beyond QRS onsets, T-wave ends and P-wave onsets as knots. As we can see from Fig. 3.16, for a sinusoidal noise frequency ranging from 0.1 Hz to 0.5 Hz, the approach based on detecting QRS onsets, T-waves ends, and P-wave onsets, and applying all these fiducial points as knots for polynomial fitting, is more accurate than the other simpler methodology.

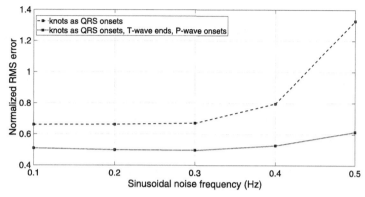

FIGURE 3.16

Performance comparison of two possible approaches for polynomial filtering for a sinusoidal noise frequency, ranging from 0.1 Hz to 0.5 Hz, taking into account only QRS onsets as knots, and considering beyond QRS onsets, T-wave ends and P-wave onsets as knots.

3.2.1.4 Wavelet Filtering

Although the family of techniques known as Wavelet filtering (or Wavelet transform) belong to the class of linear methods, because they can also be implemented as FIR filter banks, their extensive application nowadays, and robust results, justifies a detailed explanation. Wavelet transform involves combining the time-domain and frequency-domain analysis of a signal (Gari et al., 2006). This joint time-frequency analysis provides filtering of both persistent signal and short transient sources of noise. In addition to Wavelet transform, a number of time-frequency methods are available for signal analysis, such as short time Fourier transform (STFT), Wigner–Ville transform (WVT), and Choi–Williams distribution (CWD) (Gari et al., 2006). In relation to these methods, Wavelet transform appears as the most favored tool by researches as it possesses frequency-dependent windowing, which allows for arbitrary high-resolution of the high-frequency signal components.

The continuous wavelet transform (CWT) of a continuous time signal $x(t)$ is defined as

$$T(a, b) = \frac{1}{\sqrt{a}} \int_{-\infty}^{+\infty} x(t)\psi^*(\frac{t-b}{a})dt, \qquad (3.17)$$

where $\psi^*(t)$ is the complex conjugate of the mother wavelet $\psi(t)$, which is shifted by a time b and dilated or contracted by a factor a prior to computing its correlation with the signal $x(t)$.

In order to be classified as a wavelet, a function must satisfy a set of mathematical criteria (Addison, 2005):

- It must have finite energy:

$$E = \int_{-\infty}^{\infty} |\psi(t)|^2 dt < \infty; \tag{3.18}$$

- If $\hat{\phi}(f)$ is the Fourier transform of $\psi(t)$, then

$$C_g = \int_{0}^{\infty} \frac{|\hat{\phi}(f)|^2}{f} df < \infty, \tag{3.19}$$

that is, $\hat{\phi}(0) = 0$ (it must have a zero mean);
- for complex (or analytic) wavelets, the Fourier transform must both be real and vanish for negative frequencies.

As an example of a Wavelet function, the Mexican hat wavelet is the second derivative of a Gaussian function, which is computed as

$$\psi(t) = (1 - t^2)e^{-\frac{t^2}{2}}. \tag{3.20}$$

The Mexican hat wavelet has been widely applied for ECG denoising and QRS enhancing due to its similarity with the normal QRS morphology pattern. As illustrative examples of applying Continuous Wavelet Transform (CWT) over synthetic ECG signals, Figs. 3.17(A), 3.17(B), 3.17(C), and 3.17(D) present, respectively, an excerpt of synthetic ECG signal with a 0.4-Hz frequency sinusoidal noise, the same excerpt of ECG filtered by CWT (factor $a = 1$), a second version with factor $a = 2^4$, and a third version with factor $a = 2^9$. From the obtained results, we can observe the property of joint time-frequency analysis for Continuous Wavelet Transform, which allows discriminating physiological content (QRS complex, T-wave, and P-wave) and low-frequency noise.

Discrete wavelet transform

Discrete wavelet transform (DWT) consists of choosing scales and positions based on powers of two, known as dyadic scales and positions, which provides efficiency and accuracy for denoising purposes and also for enhancing specific physiologic content.

DWT employs a dyadic grid (integer power of two scaling in a and b) and orthonormal wavelet basis functions, exhibiting zero redundancy (Addison, 2005). An intuitive way to sample the parameters a and b is to use a logarithmic discretization of the factor a, and link this to the size of steps taken between b locations. This kind of discretization has the form

$$\psi_{m,n}(t) = \frac{1}{\sqrt{a_0^m}} \psi\left(\frac{t - nb_0 a_0^m}{a_0^m}\right), \tag{3.21}$$

where the integers m and n control the wavelet dilation and translation, respectively. In this case, a_0 is a specified fixed dilation step parameter, and b_0 is the location

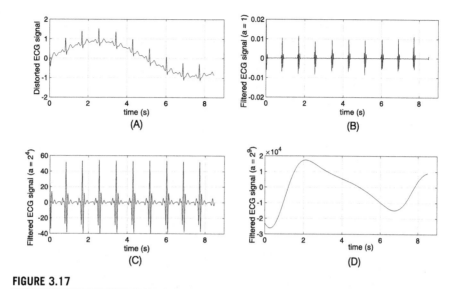

FIGURE 3.17

Applying Continuous Wavelet Transform for discriminating ECG physiological content and low-frequency noise. (A) Distorted ECG signal with a 0.4-Hz frequency sinusoidal noise. (B) ECG signal filtered by CWT, with scale $a = 1$. (C) ECG signal filtered by CWT, with scale $a = 2^4$. (D) ECG signal filtered by CWT, with scale $a = 2^9$.

parameter. A common choice for a_0 and b_0 are 2 and 1 m, respectively Addison (2005).

Substituting $a_0 = 2$ and $b_0 = 1$ into Eq. (3.21), the dyadic grid wavelet can be written as

$$\psi_{m,n}(t) = 2^{-m/2}\psi(2^{-m}t - n). \tag{3.22}$$

Using the dyadic grid wavelet referred to in Eq. (3.22), the discrete wavelet transform (DWT) is given as

$$T_{m,n} = \int_{-\infty}^{\infty} x(t)\psi_{m,n}(t)dt, \tag{3.23}$$

where $T_{m,n}$ is known as the detail coefficient at scale and location indices m, n.

Mallat (1989) developed an efficient and reliable algorithm to compute DWT decompositions, using consecutive filters and decimators, and also the inverse DWT, that is, the process of reconstructing modified wavelet coefficients, using consecutive filters and upsamplers, as illustrated in Fig. 3.18. At the different stages, the high-pass filters ($H(z)$) and low-pass filters ($G(z)$) determine the corresponding details $d_k[n]$ and approximations $a_k[n]$ coefficients, respectively AlMahamdy and Riley (2014). By using a filter bank to decompose a signal allows one to selectively examine or modify the content of a signal within chosen bands for the purpose of compression,

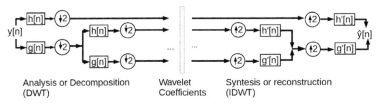

Analysis or Decomposition   Wavelet   Syntesis or reconstruction
(DWT)   Coefficients   (IDWT)

FIGURE 3.18

A synthesized schema for wavelet decomposition and reconstruction through filter banks.

filtering, or signal classification, and also reconstructing the filtered and/or enhanced wavelet coefficients.

Considering, as an illustrative example, two levels of decomposition, an input signal $y[n]$ is filtered by a high-pass filter $h[n]$ and downsampled, which provides obtaining the first set of details coefficients $d_1[n]$. The same input signal $y[n]$ is filtered by a low-pass filter $g[n]$ and downsampled, which provides obtaining the first set of approximations coefficients $a_1[n]$. Then, the sequence $a_1[n]$ is filtered by the high-pass filter $h[n]$ and again downsampled, which provides obtaining the second set of details coefficients $d_2[n]$. The same sequence is filtered by the low-pass filter $g[n]$ and downsampled, which provides obtaining the second set of approximations coefficients $a_2[n]$.

In general, the overall waveform of a signal will be primarily contained in the approximation coefficients, and short-term transients or high frequency activity will be contained in the detail coefficients. Therefore depending on the sampling frequency of the original ECG signal $y[n]$, as we eliminate the details coefficients at the various stages, and reconstruct the signal, using only approximation coefficients, we will recover the major morphological component and progressively discriminate lower and lower frequency components. At deeper decompositions, we can recover the baseline wander and, then, subtract from the original signal.

While coefficients $a_k[n]$ are related to low frequency part of the signal (main features and information), detail coefficients $d_k[n]$ are important to preserve the perfect shape of the characteristic waves when reconstruction is invoked (AlMahamdy and Riley, 2014). As $d_k[n]$ coefficients at higher levels of noiseless signals are sparse, larger coefficients in these high levels may be assumed to contain physiologic information plus noise, whereas the remaining coefficients may be considered to be pure noise. So the idea of denoising signals involves using wavelet thresholding of detail coefficients (AlMahamdy and Riley, 2014).

Therefore when performing ECG denoising, two specific tasks are required: wavelet decomposition and thresholding. Performing a wavelet decomposition involves the following steps (Gari et al., 2006):

- Select a mother-wavelet appropriate for analyzing the ECG signal. The wavelet should have morphological features, which match those to be extracted, highlighted or detected in the ECG signal;

- Determine the high-pass and low-pass decomposition filters $H_0(z)$ and $H_1(z)$ for filter bank implementation and computing of wavelet coefficients;
- Determine the high-pass and low-pass reconstruction filters $F_0(z)$ and $F_1(z)$ for filter bank implementation, so that a new version of the ECG signal can be reconstructed from the modified wavelet coefficients.

Considering the family of mother-wavelets Daubechies, named *dbk*, where *k* refers to the number of vanishing moments in the low-pass and high-pass filters, Figs. 3.19(A) to (J) illustrate a set of decomposition low-pass and high-pass filters, respectively, related to db1, db2, db4, db8, and db10. The corresponding reconstruction high-pass and low-pass filters are simply the time-reversed version of each filter. If the mother-wavelet has *n* vanishing moments, the wavelet transform may be interpreted as a multiscale differential operator of order *n* (Mallat, 1999).

As a practical application, we used an excerpt of synthetic ECG signal (McSharry et al., 2003), lasting 60 s, with a sampling frequency of 360 Hz, and apply a wavelet decomposition until the eighth level, which allow us to obtain $a_8[n]$. Considering that the group of filters divides up a signal into various spectral components, which is termed subband coding, frequency bands are divided in half by the high-pass and low-pass filters in the decomposition process, and the coefficients $a_8[n]$ provide content about the range 0 to 0,703125 Hz. The process of reconstructing baseline estimation is implemented, keeping the values of $a_8[n]$ and setting all the coefficients d_k to zero. As mother wavelet, we use DB10. Figs. 3.20(A), (B) illustrate, respectively, an excerpt of synthetic ECG signal added with a 5 Hz-sinusoidal noise (solid line), with the corresponding reconstructed baseline signal, and the resultant filtered ECG signal, after the noisy ECG signal has been subtracted from the baseline estimation.

Taking the normalized RMS error as an evaluation metric, we can compare the performance of the discrete wavelet transform, varying the frequency of the sinusoidal noise at the range [0.1–1 Hz] and applying different mother wavelets. Fig. 3.21 presents the evolution of the normalized RMS error at the considered frequency range, considering as mother-wavelets the functions db2, db3, db4, db7, and db10.

3.2.2 POWERLINE INTERFERENCE REMOVAL

The powerline interference represents a common noise source in the ECG and other physiologic signals recorded from the body surface. Depending on the country region, such noise is characterized by a 50 or 60 Hz sinusoidal interference, possibly accompanied by harmonics. This kind of noise is classified as a narrowband and hampers visual analysis of ECG and automatic segmentation processes, mainly as a result of low-amplitude waveforms, making the boundary regions of P-waves and T-waves unidentifiable (see Fig. 3.1). Several techniques and approaches have been developed and evaluated for removing such interference, as band-stop FIR and IIR filtering, adaptive filters, subtraction of estimated components, wavelet transform, and advanced techniques, which handle variations in power line frequency (Sörnmo and Laguna, 2005).

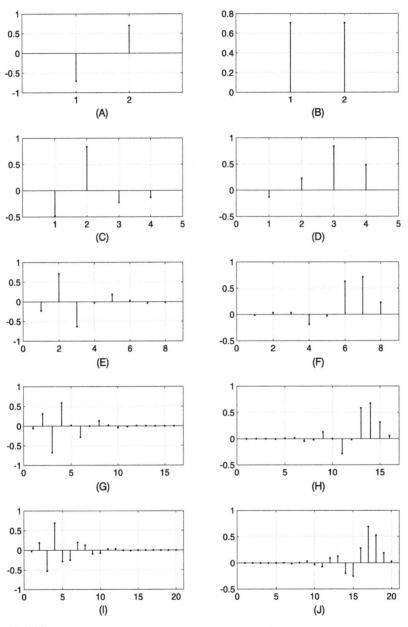

FIGURE 3.19

Set of high-pass and low-pass decomposition filters, respectively, related to wavelets daubechies: db1 (A) and (B), db2 (C) and (D), db4 (E) and (F), db8 (G) and (H), and db10 (I) and (J).

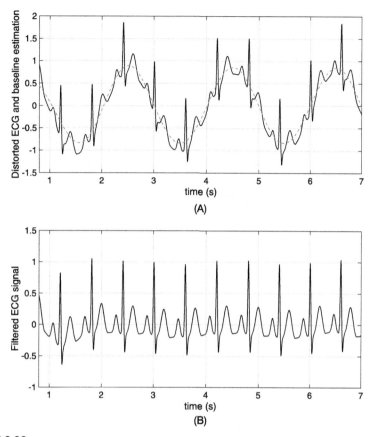

FIGURE 3.20

Filtering based on discrete wavelet transform: (A) an excerpt of a 6-s synthetic ECG signal with a 0.5-Hz frequency sinusoidal noise (solid line) and its baseline estimation (dashed line), obtained through wavelet decomposition until the eighth level, and (B) the resultant subtracted ECG signal.

3.2.2.1 Linear Filtering

If we consider a filter defined by a complex-conjugated pair of zeros that lie on the unit circle, at the interfering frequency ω_0, i.e., $z_{1,2} = e^{\pm j\omega_0}$, we can obtain the simplest approach for removing powerline interference, which is based on a second-order FIR filter (Sörnmo and Laguna, 2005).

This FIR filter has the transfer function

$$H(z) = (1 - e^{j\omega_0}z^{-1})(1 - e^{-j\omega_0}z^{-1}). \tag{3.24}$$

This filter represents a notch with a relatively large bandwidth. This means that it will attenuate not only the powerline frequency but also the ECG content with frequencies close to ω_0 (Sörnmo and Laguna, 2005).

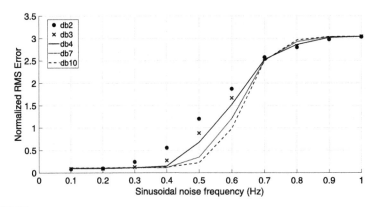

FIGURE 3.21

Performance comparison of different mother-wavelets from the family Daubechies, by varying the frequency of a sinusoidal noise from 0.1 to 1 Hz.

A possible modification aiming to increase the performance of the FIR filter, that is, to make the notch become more selective, is to introduce a pair of complex-conjugated poles positioned at the same angle as the zeros $z_{1,2}$, but at a radius r (Sörnmo and Laguna, 2005)

$$p_{1,2} = re^{\pm j\omega_0}, \tag{3.25}$$

where $0 < r < 1$.

The transfer function of the resulting IIR filter is given by

$$H(z) = \frac{1 - 2\cos(\omega_0)z^{-1} + z^{-2}}{1 - 2r\cos(\omega_0)z^{-1} + r^2 z^{-2}}. \tag{3.26}$$

The bandwidth of the notch is directly influenced by the pole radius and it becomes more selective as radius r approaches the unity value, that is, the unit circle. Figs. 3.22(A)–(D) show magnitude responses for four second-order IIR filters, which represent a notch for the 60-Hz interfering frequency, whose zeros are positioned at $e^{\pm j\omega_0}$ and poles are positioned at $e^{r.\pm j\omega_0}$, with $\omega_0 = (60/360)\pi$, considering a 360-Hz sampling frequency, and r presents the possible values 0.75, 0.85, 0.95, and 0.99.

Figs. 3.22(A)–(D) demonstrate that despite the narrowband of the notch filters decrease, with the increasing of r, the degree of the magnitude attenuation at the interfering frequency also decreases. One more important fact hereto, bearing on the time-domain, is that increasing the parameter r also causes a ringing artifact in the output signal, which is due to the increased transient response time of the filter. As a computing experiment, we synthesized an excerpt of ECG signal using the ECG generator proposed by McSharry et al. (2003), lasting 60 s, with a 360-Hz sampling frequency, introduced a 60-Hz artificial sinusoidal noise and applied a

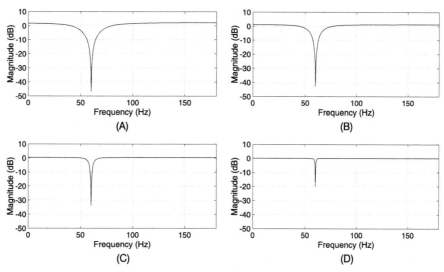

FIGURE 3.22

Magnitude response for four second-order IIR filters, which represent a notch for the 60-Hz interfering frequency, whose zeros are positioned at $e^{\pm j\omega_0}$ and poles are positioned at $re^{\pm j\omega_0}$, with $\omega_0 = (60/360)\pi$, considering a 360-Hz sampling frequency, and $r = 0.75$ (A), 0.85 (B), 0.95 (C), and 0.99 (D).

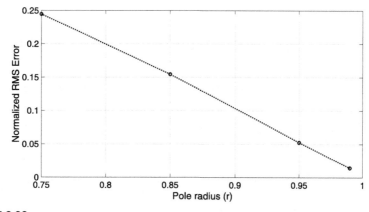

FIGURE 3.23

Evolution of the normalized RMS error as the pole radius increases.

second-order notch IIR filters with the values for pole radius $r = 0.75$, 0.85, 0.95, and 0.99. The evolution of the normalized RMS error between the original "clean" signal and the resultant signal filtered by the corresponding IIR filters is illustrated by Fig. 3.23, where we can observe a general performance increase as the parameter r increases.

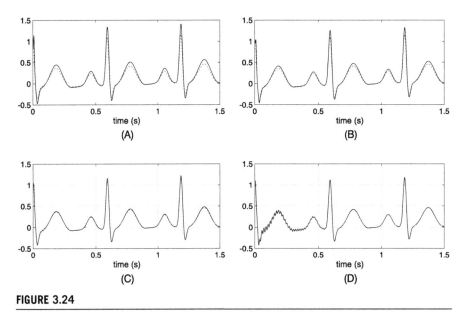

FIGURE 3.24

Excerpt of filtered ECG signals after applying second-order notch IIR filters (solid line) and the reference "clean" ECG signal (dashed line). (A) Filtered ECG signal ($r = 0.75$). (B) Filtered ECG signal ($r = 0.85$). (C) Filtered ECG signal ($r = 0.95$). (D) Filtered ECG signal ($r = 0.99$).

Figs. 3.24(A)–(D) show can an increasing agreement between the clean reference ECG signal and the resultant filtered signal related to increasing the parameter r, but with the expense of an increasing ringing effect during the first beats, which significantly modify its diagnostic content.

Other approaches allow us to design more sophisticated and efficient filters for removing the power line interference (also including the harmonics). If we use higher-order IIR or FIR filters to obtain a narrower notch or by employing, we will always face the ringing effect as an artifact over the QRS complexes, since increased frequency resolution is obtained at the detriment of decreased time resolution.

As an example, Fig. 3.25 illustrates the result of applying a band-stop FIR filtering with 1500 coefficients (Hamming window), cutoff frequencies of 59 Hz and 61 Hz, over a synthetic ECG signal with a 360-Hz sampling frequency. We can observe the ringing effect over the ST-segment, and also over the T-wave.

Some approaches are based on the idea of subtracting a sinusoid, generated internally by a filter from the observed signal, such that its amplitude are adapted to the powerline interference present in the observed digitized ECG signal. Since the real amplitude of the interference is unknown and changing with time, it is preferable to generate the sinusoid recursively by adapting their amplitudes at every sample (Sörnmo and Laguna, 2005).

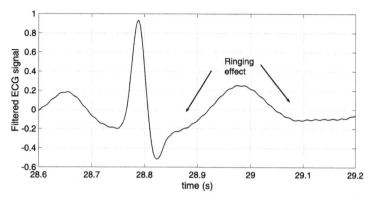

FIGURE 3.25

Result of applying a band-stop FIR filtering for removing 60-Hz sinusoidal, considering 1500 coefficients (Hamming window), cutoff frequencies of 59 Hz and 61 Hz, over a synthetic ECG signal with a 360-Hz sampling frequency.

So given an estimated and synthesized sinusoid $v[n]$, their amplitudes can be generated by an oscillator, whose transfer function is (Sörnmo and Laguna, 2005)

$$H(z) = \frac{V(z)}{U(z)} = \frac{1}{1 - 2\cos(\omega_0)z^{-1} + z^{-2}}, \tag{3.27}$$

and the corresponding difference equation is (Sörnmo and Laguna, 2005)

$$v[n] = 2\cos\omega_0 v[n-1] - v[n-2] + u[n], \tag{3.28}$$

using the initial conditions $v[-1] = v[-2] = 0$, and $u[n]$ as the unitary impulse function $\delta[n]$.

An error function and its first regressive difference are computed for indication, and evaluating the quality of the prediction of the powerline interference contained in the ECG signal $x[n]$, given as

$$e[n] = x[n] - v[n],$$
$$e'[n] = e[n] - e[n-1] = x[n] - x[n-1] - (v[n] - v[n-1]). \tag{3.29}$$

The reason for using the first difference is to make the filter insensitive to the DC level. Depending on the sign of $e'[n]$, the current value of $v[n]$ is adapted by a positive or negative increment α (Sörnmo and Laguna, 2005),

$$\hat{v}[n] = v[n] + \alpha \, sgn(e'[n]), \tag{3.30}$$

where $sgn(x)$ refers to the sign function.

Finally, the output signal of the nonlinear filter results from subtraction of the estimate $\hat{v}[n]$ for the powerline interference from the original ECG signal $x[n]$

$$y[n] = x[n] - \hat{v}[n]. \tag{3.31}$$

By applying computing simulations with synthetic ECG signals, we can observe that too small a value of α causes the filter to require more time (samples) to converge to a noiseless reference signal, whereas too large a value of α causes the filter to be faster at the convergence process, but also introduces extra noise in the filtered signal, because of the large step alterations, which will occur in the estimated noise amplitude $\hat{v}[n]$.

In Fig. 3.26(A)–(C), we illustrate the convergence process of the above detailed nonlinear filter, considering, respectively, $\alpha = 0.05$, 0.5, and 5 µV, for removing a 60-Hz sinusoidal noise from an excerpt of ECG signal lasting 65 s. We can observe the failure of the nonlinear filter for removing the sinusoidal noise around the instant 61.5 s, when $\alpha = 0.05$ µV. For $\alpha = 0.5$ µV, the filtering process behaves successfully around the same instant, whereas for $\alpha = 5$ µV, an extra noise is introduced in the signal.

Considering an excerpt of a real 7.5-s ECG signal (author's ECG) collected through the shield Olimex board and model SHIELD EKG-EMG, we applied the referred to non-linear filter, aiming to detect the actual frequency peaks related to powerline frequency as a time-domain approach. Therefore we varied the relative angular frequency ω_0 among the corresponding values related to the interval [120–124 Hz] and [60–64 Hz], with a step equal to 0.01 Hz. For each frequency value of the parameter ω_0, we computed the standard deviation of the filtered signal, i.e., $y[n]$, using $\alpha = 0.01$. The evolution of the standard deviation values can be analyzed through Figs. 3.27(A), (B), respectively, for searches around the frequencies 120 Hz and 60 Hz. The minimum values for standard deviations were obtained, respectively, for frequencies 122.59 Hz and 61.36 Hz, which can be validated by analyzing the magnitude frequency response of the original signal (see Fig. 3.27(C)). Then we applied the nonlinear filter using ω_0 values, respectively, equal to $122.59/128 \cdot \pi = 3.0058$ and $61.36/128 \cdot \pi = 1.5060$ for obtaining the filtered signal (see Fig. 3.27(D)).

Other possible strategy for eliminating powerline interference is the combination of different filters, as for example the notch second-order IIR filters, with discrete wavelet transforms (DWT). Considering, as explained above, that filter banks related to DWT divide up a signal into various spectral components, which is termed subband coding, and frequency bands are divided in half by the high-pass and low-pass filters in the decomposition process, we can apply one level of decomposition and take just the first approximation (coefficients $a_1[n]$) for the reconstruction process. Therefore applying, firstly, notch IIR filters allows eliminating sinusoidal frequency noise with frequency peak around 60 Hz, and applying a 1-level wavelet decomposition and reconstruction allows us to remove harmonics and other spurious noise.

Considering the same excerpt of a real 7.5-s ECG signal (author's ECG) detailed above, whose spectral content is illustrated in Fig. 3.27(C), we applied successively a notch second-order IIR filter (with parameter radius equal to 0.99) and a 1-level wavelet decomposition and reconstruction, using db7 as mother-wavelet.

Fig. 3.28(A) shows the spectral magnitude of the resultant filtered signal, and in Fig. 3.28(B), we observe the filtered ECG signal, superimposed with the original signal.

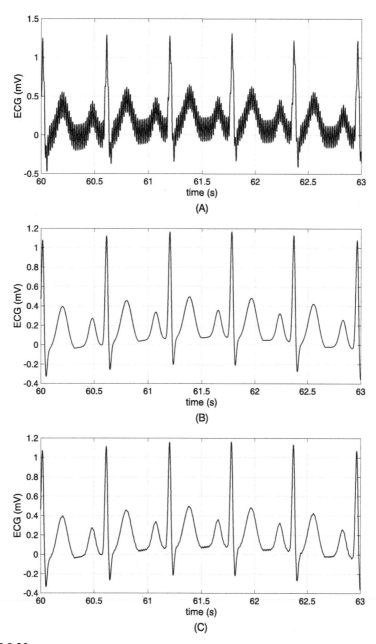

FIGURE 3.26

Convergence process of the nonlinear filter, considering $\alpha = 0.05$ (A), 0.5 (B) and 5 µV (C), for removing a 60-Hz sinusoidal noise from an excerpt of synthetic ECG signal lasting 65 s.

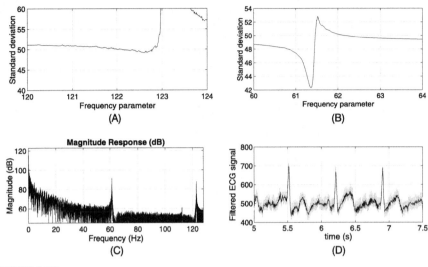

FIGURE 3.27

Applying the nonlinear filter for identifying frequency peaks related to powerline interference: (A) evolution of the standard deviation of the filtered signal for ω_0, varying throughout the values corresponding to the interval [120–124 Hz]; (B) evolution of the standard deviation of the filtered signal for ω_0, varying throughout the values corresponding to the interval [60–64 Hz]; (C) spectral magnitude of the original signal, and (D) filtered signal considering the identified frequency peaks.

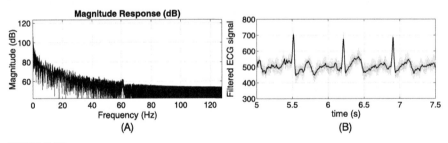

FIGURE 3.28

Combination of filtering techniques for removing power line interference (including harmonics), notch IIR filter and discrete wavelet transform: (A) spectral magnitude content of the filtered signal, and (B) resultant denoised ECG signal.

3.2.3 MUSCLE NOISE FILTERING

As already pointed out, electromyographic (EMG) noise refers to electrical activity due to muscle contractions, and its spectral content covers between DC and 10,000 Hz. In contrast to baseline wander and powerline interference, muscle noise is not removed by narrowband filtering, but represents a much more difficult challenge since the spectral content of muscle activity considerably overlaps that of the ECG characteristic waves: QRS complex, P- and T-waves. Beat averaging and median filtering are widely used estimators in determining the dominant beat morphology, given that beat averaging is expected to provide efficient suppression of muscle noise, whereas median filtering of beats is known to be less sensitive to outliers, such as sudden cardiac baseline shifts, spikes or abnormal beats (Gari et al., 2006; Jager et al., 1991).

In beat averaging, beats are time aligned by the fiducial point of each QRS complex (in general, the R-wave) and added together. Assuming that the noise is uncorrelated with the physiologic ECG signal, and it is uncorrelated with itself on a beat-to-beat analysis, the expected signal-to-noise ratio (SNR) of the average beat is improved for $\sqrt{Q(j)}$, where $Q(j)$ is the number of beats included into the jth average (Gari et al., 2006).

Despite providing a robust analysis of the predominant beat morphology, successful noise reduction by ensemble averaging is restricted to one particular QRS morphology at a time, and requires that several beats be available. Considering the different limitations from several filtering techniques, it is assumed that there is still a need to develop signal processing techniques, which can efficiently attenuate muscle noise.

A robust and alternative technique for attenuation of muscle noise and other types of ECG noise is the empirical mode decomposition (EMD) (Wu and Huang, 2009). The aim of the EMD is to decompose the signal into a sum of intrinsic mode functions (IMFs), where an IMF is defined as a function with equal number of extrema and zero crossings with its envelopes (Blanco-Velasco et al., 2008).

Given an original ECG signal $x[n]$, the EMD method starts by identifying all the local maxima and minima. All the local maxima are connected by a cubic spline, composing the upper envelope $e_u[n]$. The local minima are also connected by a spline curve composing the lower envelope $e_l[n]$. Denoting the mean of the two envelopes as $m_1[n]$, the first proto-IMF $h_1[n]$ is obtained as (Blanco-Velasco et al., 2008)

$$h_1[n] = x[n] - m_1[n]. \tag{3.32}$$

Since $h_1[n]$ still contains multiple extrema in between zero crossings, the above process is performed again, and applied repetitively to functions $h_k[n]$ until the first IMF $c_1[n]$ is obtained. A stop criteria is applied, which is related to the sum of difference (SD) (Blanco-Velasco et al., 2008)

$$SD = \sum_{t=0}^{T} \frac{|h_{k-1}[n] - h_k[n]|^2}{h_{k-1}^2[n]}, \tag{3.33}$$

a measure smaller than a threshold. When this condition is verified, the first IMF $c_1[n]$ is obtained as

$$r_1[n] = x[n] - c_1[n]. \tag{3.34}$$

The signal $r_1[n]$ represents a residue and still contains useful information. Treating the residue as a new signal and applying the above procedure allow us to obtain (Blanco-Velasco et al., 2008)

$$r_2[n] = r_1[n] - c_2[n], \tag{3.35}$$

$$r_N[n] = r_{N-1}[n] - c_N[n]. \tag{3.36}$$

The procedure stops when the residue $r_N[n]$ is either a constant, a monotonic slope, or a function with only one extremum. Combining all the above equations yields the EMD of the original signal

$$x[n] = \sum_{k=1}^{N} c_k[n] + r_N[n]. \tag{3.37}$$

The result of empirical mode decomposition produces N intrinsic mode functions and a residue signal, which is in general a monotonic slope. Referring to $c_n(t)$ as the nth-order IMF, we can verify that lower-order IMFs capture fast oscillation modes, whereas higher-order IMFs typically represent slow oscillation modes. Noise components encountered in ECG signal processing applications lie in the first IMFs, with the exception of baseline wandering. However, although most ECG signal power is concentrated in lower frequencies, QRS complex spreads across mid and high frequency bands. Therefore QRS complex content spreads also over the lower-order IMFs, which implies that simply removing lower-order IMFs, and partially reconstructing the signal, introduce several QRS complex distortion.

Considering an excerpt of a real 5-s ECG signal (author's ECG) collected through the shield Olimex board and model SHIELD EKG-EMG, we apply the above described algorithm for obtaining EMD, which results in ten IMFs. Fig. 3.29 presents the original ECG (at bottom), which contains powerline interference, baseline wander, and also muscle activity noise, and part of resulting IMF 1 to 7.

Fig. 3.29 demonstrates that the first and second IMFs contain strong noise components, and the oscillatory patterns of the QRS complex become more apparent, starting from the third IMF.

Blanco-Velasco et al. (2008) developed an approach for ECG denoising, considering high-frequency muscle activity noise, powerline interference, and also baseline wandering, based on EMD and partial reconstruction. The QRS complex spreads across mid- and high-frequency bands and have the following methodology proposes: (a) delineate and separate the QRS complex; (b) use proper windowing to preserve the QRS complex; (c) use statistical tests to determine the number of IMFs contributing to the noise, and (d) filter the noise by partial reconstruction.

Fig. 3.30 illustrates the process of denoising through empirical mode decomposition for the same excerpt of a real 5-s ECG signal (author's ECG), collected through the shield Olimex board and model SHIELD EKG-EMG. This approach identifies the first three IMFs considered to be noise components (mid and high noise frequencies),

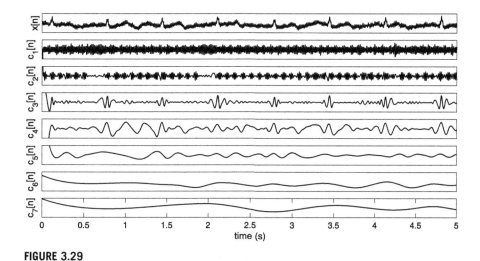

FIGURE 3.29

Empirical mode decomposition of a real noisy 5-s ECG signal. From top to bottom: original ECG and part of resulting IMF 1–7.

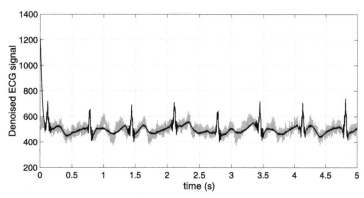

FIGURE 3.30

Denoised ECG signal through applying EMD decomposition, window functions to preserve QRS and attenuation of noise components in the first three IMFs.

and applies window functions to these IMFs to preserve QRS complex. The sum of windowed IMFs, the remaining IMFs, and the residues form the reconstructed signal.

Furthermore, discrete wavelet transform is also applied for attenuating muscle noise filtering. The approach developed by Ardhapurkar et al. (2012) presents a denoising method based on the modeling of discrete wavelet coefficients of ECG in selected sub-bands, with Kernel density estimation. The authors consider that the wavelet domain choice of wavelet function depends on the application and the shape of a signal. After an evaluation of the Mean Square Error (MSE) between the original signal and the reconstructed signal on several noise-corrupted records, with 500-Hz sampling rate, the Daubechies wavelet of order 4 (db4) is selected. As the selection of the decomposition levels depends on the sampling frequency and the spectral content of the signal, their approach proposes:

- Setting coefficients cA10 and cD10 to zero to reduce baseline wandering;
- Setting detail coefficients cD1 and cD2 to zero to eliminate high-frequency noise;
- Clipping off detail coefficients cD8 and cD9 to reduce motion artifacts;
- Applying kernel density estimation and threshold to detail coefficients cD3 and cD4 to reduce interference of muscle activity noise and also power line;
- Retaining detail coefficients cD5 to cD7 carrying clinical information.

3.3 CONCLUSIONS

In this chapter, we reviewed the theory concerned with the spectral content of the ECG signal and how the noise spectra impacts ECG content, important concepts as signal-to-noise ratio (SNR), magnitude, phase response, and power spectrum of a signal. We emphasized that ECG filtering covers three specific applications: baseline wander suppression, line frequency rejection, and muscle artifact reduction. Then we described and implemented, through computing simulations, several techniques based on linear and time-variant filtering, polynomial fitting, wavelet filtering and empirical mode decomposition. It is clear that some classic filtering techniques introduce significant distortion within the filtered ECG signal, are poorly inefficient, and unable to remove simple noise classes, such as the baseline wandering. Some emphasis should be given to Wavelet transform and empirical mode decomposition, which block the most common types of noise with a minimum distortion. However, depending on the final application, a certain level of distortion may be tolerated. Therefore the most appropriate filtering technique to be applied depends on what we want to extract from the ECG signal. Despite the considerable ECG signal denoising evolution, a high research demand for the analysis of the impact of different filtering techniques over specific tasks within ECG signal processing, such as wave segmentation, feature extraction, and arrhythmia classification is apparent.

REFERENCES

Addison, P.S., 2005. Wavelet transforms and the ECG: a review. Physiological Measurement 26 (5), R155.

AlMahamdy, M., Riley, H.B., 2014. Performance study of different denoising methods for ECG signals. Procedia Computer Science 37, 325–332.

Ardhapurkar, S., Manthalkar, R., Gajre, S., 2012. ECG denoising by modeling wavelet sub-band coefficients using kernel density estimation. Journal of Information Processing Systems 8 (4), 669–684.

Bailey, J.J., Berson, A.S., Garson Jr., A., Horan, L.G., Macfarlane, P.W., Mortara, D.W., Zywietz, C., 1990. Recommendations for standardization and specifications in automated electrocardiography: bandwidth and digital signal processing. A report for health professionals by an ad hoc writing group of the committee on electrocardiography and cardiac electrophysiology of the Council on Clinical Cardiology, American Heart Association. Circulation 81 (2), 730.

Berson, A.S., Pipberger, H.V., 1966. The low-frequency response of electrocardiographs, a frequent source of recording errors. American Heart Journal 71 (6), 779–789.

Blanco-Velasco, M., Weng, B., Barner, K.E., 2008. ECG signal denoising and baseline wander correction based on the empirical mode decomposition. Computers in Biology and Medicine 38 (1), 1–13.

De Boor, C., 1978. A Practical Guide to Splines. Applied Mathematical Sciences, vol. 27. Springer-Verlag, New York.

Friesen, G.M., Jannett, T.C., Jadallah, M.A., Yates, S.L., Quint, S.R., Nagle, H.T., 1990. A comparison of the noise sensitivity of nine QRS detection algorithms. IEEE Transactions on Biomedical Engineering 37 (1), 85–98.

Gari, D.C., Francisco, A., Patrick, E., 2006. Advanced Methods and Tools for ECG Data Analysis. Artech House, Inc.

Goldberger, A.L., Amaral, L.A., Glass, L., Hausdorff, J.M., Ivanov, P.C., Mark, R.G., Mietus, J.E., Moody, G.B., Peng, C.K., Stanley, H.E., 2000. PhysioBank, PhysioToolkit, and PhysioNet. Circulation 101 (23), 215–220.

Jager, F., Mark, R., Moody, G., 1991. Analysis of transient st segment changes during ambulatory monitoring. In: Computers in Cardiology 1991, Proceedings. IEEE, pp. 453–456.

Lemay, M., Jacquemet, V., Forclaz, A., Vesin, J., Kappenberger, L., 2005. Spatiotemporal QRST cancellation method using separate QRS and T-waves templates. In: Computers in Cardiology. IEEE, pp. 611–614.

Lugovaya, T.S., 2005. Biometric Human Identification Based on ECG. PhysioNet.

Madeiro, J.P., Cortez, P.C., Marques, J.A., Seisdedos, C.R., Sobrinho, C.R., 2012. An innovative approach of QRS segmentation based on first-derivative, Hilbert and Wavelet Transforms. Medical Engineering & Physics 34 (9), 1236–1246.

Mallat, S.G., 1989. A theory for multiresolution signal decomposition: the wavelet representation. IEEE Transactions on Pattern Analysis and Machine Intelligence 11 (7), 674–693.

Mallat, S., 1999. A Wavelet Tour of Signal Processing. Academic Press.

McSharry, P.E., Clifford, G.D., Tarassenko, L., Smith, L.A., 2003. A dynamical model for generating synthetic electrocardiogram signals. IEEE Transactions on Biomedical Engineering 50 (3), 289–294.

Murthy, V.K., Grove, T.M., Harvey, G.A., Haywood, L.J., 1978. Clinical usefulness of ECG frequency spectrum analysis. In: Proceedings of the Second Annual Symposium on Computer Application in Medical Care. 1978. IEEE, pp. 610–612.

Oppenheim, A.V., 1999. Discrete-Time Signal Processing. Pearson Education, India.

Pipberger, H., Arzbaecher, R., Berson, A., 1975. American heart association committee on electrocardiography: recommendations for standardization of leads and of specifications for instruments in electrocardiography and vectorcardiography. Circulation 52, 11–31.

Semmlow, J.L., Griffel, B., 2014. Biosignal and Medical Image Processing. CRC Press.

Singh, A., 1987. Low-frequency requirements for electrocardiographic recording of ST segments. American Journal of Cardiology 60 (10), 939.

Sörnmo, L., Laguna, P., 2005. Bioelectrical Signal Processing in Cardiac and Neurological Applications, vol. 8. Academic Press.

Thakor, N.V., Webster, J.G., Tompkins, W.J., 1984. Estimation of QRS complex power spectra for design of a QRS filter. IEEE Transactions on Biomedical Engineering (11), 702–706.

Tompkins, W.J., 1993. Biomedical Digital Signal Processing: C-Language Examples and Laboratory Experiments for the IBM PC. Hauptbd. Prentice Hall.

Villarrubia, G., De Paz, J.F., Bajo, J., Corchado, J.M., 2014. EKG mobile. Advanced Science and Technology Letters 49, 95–100.

Welch, P., 1967. The use of fast Fourier transform for the estimation of power spectra: a method based on time averaging over short, modified periodograms. IEEE Transactions on Audio and Electroacoustics 15 (2), 70–73.

Wu, Z., Huang, N.E., 2009. Ensemble empirical mode decomposition: a noise-assisted data analysis method. Advances in Adaptive Data Analysis 1 (01), 1–41.

Techniques for QRS Complex Detection

4

João Paulo do Vale Madeiro*, José Maria da Silva Monteiro Filho†,
Priscila Rocha Ferreira Rodrigues†

*Institute for Engineering and Sustainable Development (IEDS), University for the International Integration of the Afro-Brazilian Lusophony – UNILAB, Redenção, Ceará, Brazil
†Department of Computing Science, Federal University of Ceara, Fortaleza, Ceará, Brazil

4.1 CLASSICAL STRUCTURE FOR QRS DETECTION

The QRS complex is commonly the most expressive waveform of the ECG signal, considering the aspects of amplitude and period of oscillation. We may say that, in general, the high and/or steep amplitude of the R-wave makes the task of QRS automatic detection more easily than for the other characteristic waves, thus consisting of the first step for the complete segmentation of the ECG signal. However, depending on morphological and physiological aspects, occurrence of cardiac diseases, and also noise incidence, an accurate automatic detection of QRS may be challenging. Thus considering multiple situations related to signal acquisition and patient conditions, a QRS detection algorithm must be able to detect a large number of different QRS morphologies in order to be clinically feasible and able to recognize sudden or gradual changes of the predominant QRS morphology and/or the prevailing heart rate rhythm.

Furthermore, the correct and accurate detection and delineation of the QRS complex are fundamental conditions for detection and segmentation of the other waves (P- and T-waves), and also serve as basis for the construction of algorithms for automatic recognition of cardiac arrhythmia patterns. Also, automatic detection of the QRS complex is required for an efficient extraction of beat-to-beat intervals (RR time-series) from long electrocardiogram recordings. Accuracy of the RR series is crucial for a reliable heart rate variability analysis, which provides a quantitative assessment of cardiac-autonomic function in health and disease states (Arzeno et al., 2008).

Noise and interference are inherent within the process of ECG signal acquisition, as extensively detailed in the previous chapter. However, the term "noise" acquires a wider meaning when considered from a QRS-detection point of view. Thus P- and T-waves, although being part of the physiological content of the ECG signal, must be treated as noise for QRS detection. Actually, in addition to noise, other signal issues or problems may affect the QRS-detection accuracy. Sörnmo and Laguna (2005) enumerate two categories of signal and noise problems:

Developments and Applications for ECG Signal Processing. https://doi.org/10.1016/B978-0-12-814035-2.00010-4

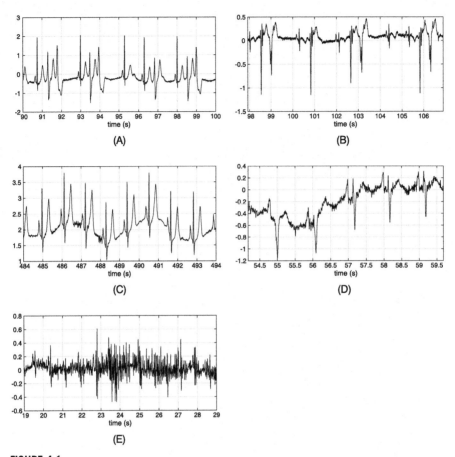

FIGURE 4.1

Types of issues and noise in the ECG from a QRS-detection point of view. (A) QRS morphology alternance. (B) QRS morphology and amplitude changes. (C) High-amplitude T-waves. (D) High-amplitude P-waves. (E) Burst of noise of muscular origin and artifacts similar to QRS.

1. Changes in QRS morphology of physiological origin, or changes due to artifacts;
2. Occurrence of noise: large P- or T-waves, muscular activity, or transient artifacts.

Excerpts of ECG signal illustrating QRS morphology and amplitude changes, T-waves, which could be misinterpreted as QRS complexes, burst of noise of muscular origin, and artifacts similar to QRS are presented by Fig. 4.1. One should emphasize that it may be necessary, considering ECG-related aim, to exclude episodes containing excessive noise from clinical analysis.

Considering the wide diversity of "noise" from a QRS-detection point of view, Fig. 4.2 presents a generalized framework for QRS detection widely considered by many techniques, which is based on a preprocessing stage, including linear and non-

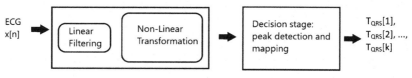

FIGURE 4.2

Common double-stage structure for QRS detectors: the input is the digitized ECG signal and the output is the occurrence times of each QRS complex.

linear filtering, and a decision stage, including peak detection and a decision logic. The input is the digitized ECG signal and the output is the number of times each QRS complex occurs. It should be emphasized that most QRS detectors described in the literature has been developed from experimental insight, example, by testing and adjusting approaches according to a wide diversity of QRS morphologies and noise incidence.

Regarding linear transformation, the detectors filter the ECG signal with a band-pass filter (or a matched filter) to suppress the P- and T-waves, and noise. At this step, waveform distortion is not a critical restriction. The focus is improving the signal-to-noise ratio to achieve a satisfactory detector performance. Concerning the passband, the center frequency of the filters varies from 10 to 25 Hz, and the bandwidth from 5 to 10 Hz (Sörnmo and Laguna, 2005).

Then the signal will be passed through a nonlinear transformation, for example, square, to enhance the QRS complexes and transform each QRS complex into a single positive peak, better suited for peak detection at decision stage. Finally, decision rules are used to determine whether QRS complexes are present in the signal. The decision rules may include adaptive threshold techniques and additional tests aiming to prevent noise influence, such as transient artifacts similar to QRS, tall T-waves, and amplitude changes.

As will be described later, based on the dual-stage framework, two main problems affect the techniques: the frequency band of the QRS signal is different for different subjects and even for different beats of the same subject; and the overlapping of frequency bands of the noise and the QRS complex. In this context, preprocessing techniques have been developed using variable filtering parameters, either by obtaining several filtered versions of the original signal, or through an adaptive algorithm for choosing the appropriate parameters for a given QRS morphology (Martínez et al., 2004; Madeiro et al., 2012).

Concerning decision stage, it is important to emphasize that QRS detection algorithms are designed to detect heartbeat occurrences, which may be understood as any sample pertaining to QRS waveform. They generally do not produce occurrence times of the QRS complexes with high temporal resolution (e.g. the peak time of the R-waves or the highest peak). Hence, it is important to provide a stage of mapping or time alignment of the detected beats, which should concern the QRS morphology classes. In this context, Lugovaya (2005) suggests twenty possible classes for QRS morphologies, each with a different number of critical (maxima or minima) points:

qR, qRs, Rs, R, RS, rSR', rR', qrSr', RSr', rR's, rS, rSr', Qr, QS, QR, qrS, qS, rsR's', QRs, and Qrs. An accurate QRS detection with high temporal resolution is essential for automatic computing of heart rate variability.

4.2 PREPROCESSING STAGE

We will now describe and compare some approaches for both ECG linear filtering and non-linear transformation, which are applied for emphasizing QRS complex and attenuating physiological origin and technical noise.

In this sense, Madeiro et al. (2012) has proposed an approach, where the full pre-processing stage is removed and replaced by a training stage, which requires an initial 10-s ECG signal excerpt. A filtering process is applied over this training interval, consisting of a filter cascade composed by Continuous Wavelet Transform and Hilbert transforms and first-derivative filter. As we have already explained in the previous chapter, the continuous wavelet transform (CWT) of a continuous time signal $x(t)$ is defined as

$$T(a,b) = \frac{1}{\sqrt{a}} \int_{-\infty}^{+\infty} x(t)\psi^*(\frac{t-b}{a})dt. \tag{4.1}$$

In this approach, the authors used the Mexican-hat wavelet as the prototype wavelet $\Psi(t)$ for both QRS detection and delineation, which is the second derivative of a Gaussian function, given by

$$Psi(t) = \frac{1}{\sqrt{2\pi}}(1 - t^2)exp(\frac{-t^2}{2}), \tag{4.2}$$

where $\psi^*(t)$ is the complex conjugate of the mother wavelet $\psi(t)$, which is shifted by a time b, and dilated or contracted by a factor a prior to computing its correlation with the signal $x(t)$.

The great similarity between the Mexican-hat and the regular morphology of a QRS complex justified for the authors the choice of that mother-wavelet.

Considering that Continuous Wavelet Transform (CWT) presents the scale factor as a variable parameter, the approach tests four different scale factors (2^0, 2^1, 2^2, and 2^3). Figs. 4.3(A)–(E) present the magnitude frequency response of four different scale factors. As we may observe, as we increase the scale factor, the passband becomes more selective at a lower center frequency.

It should be emphasized that, depending on the sampling frequency, the set of scale factors applied for QRS enhancing and denoising should be adjusted, such that high-frequency noise and low-frequency noise (including P-waves and T-waves) be attenuated.

The Hilbert transform $g(t)$ of $f(t)$ is given as

$$g(t) = \frac{1}{\pi}P\int_{-\infty}^{\infty}\frac{f(\tau)}{t-\tau}d\tau \tag{4.3}$$

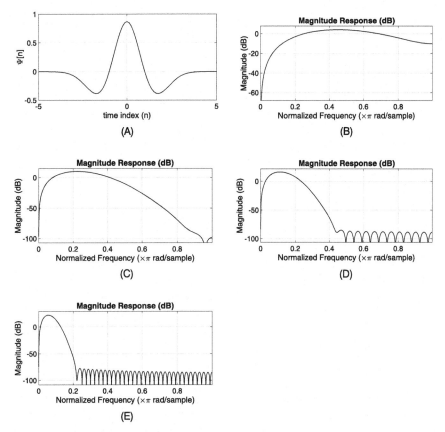

FIGURE 4.3

Evolution of the magnitude frequency response for Mexican-hat wavelet as we increase the scale factor. (A) Mexican-hat wavelet function. (B) Magnitude frequency response for scale factor equal to 2^0. (C) Magnitude frequency response for scale factor equal to 2^1. (D) Magnitude frequency response for scale factor equal to 2^2. (E) Magnitude frequency response for scale factor equal to 2^3.

when the integral exists. The P in front of the integral denotes the Cauchy principal value, which expands the class of functions for which the integral in definition exists (Hahn, 1996).

In the frequency domain, the signal is transformed with a filter of response

$$H(e^{j\omega}) = \begin{cases} -j, 0 < \omega < \pi, \\ j, -\pi < \omega < 0. \end{cases} \qquad (4.4)$$

The effects of the Hilbert transform have been explained in terms of its odd symmetry property and envelope signal. If it is applied directly over the ECG signal or over a band-pass filtered version, the QRS fiducial point is associated with a zero-crossing on the Hilbert transformed version. If it is applied on the differenti-

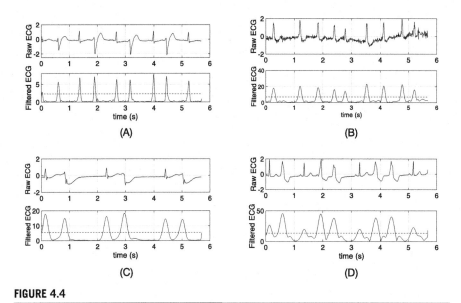

FIGURE 4.4

Undertaking preprocessing stage for ECG signal excerpts from MIT–BIH Arrhythmia Database containing different morphologies. (A) ECG 200, MIT–BIH Arrhythmia Datase: 8 QRS occurrences. (B) ECG 203, MIT–BIH Arrhythmia Datase: 9 QRS occurrences. (C) ECG 207, MIT–BIH Arrhythmia Datase: 6 QRS occurrences. (D) ECG 208, MIT–BIH Arrhythmia Datase: 9 QRS occurrences.

ated ECG, the R-peaks will be represented as peaks in the output of the transform. The Hilbert transform's all-pass characteristic prevents unnecessary signal distortion, since it shifts $-90°$ for positive and $+90°$ for negative frequencies.

The approach proposed by Madeiro et al. (2012) also applies first-derivative filter in order to emphasize the segments of the band-pass filtered signal with rapid transients, that is, the QRS complex. In discrete time, differentiation can be approximated by a filter $H(z)$, which produces a difference between successive samples (Sörnmo and Laguna, 2005),

$$H(z) = 1 - z^{-1}. \tag{4.5}$$

Concerning the referred to approach, after applying CWT over the raw ECG signal interval, a forward differentiation stage is implemented as

$$y[n] = x[n+1] - x[n]. \tag{4.6}$$

After applying the first-derivative function, the Hilbert transform is used to obtain the related analytical signal and its squared amplitude as a non-linear transformation. The squaring function leaves QRS complexes as outstanding positive peaks in the signal.

Figs. 4.4(A)–(D) illustrate excerpts of real ECG signals pertaining to MIT–BIH Arrhythmia database with their corresponding filtered versions. We also illustrate, as dashed lines, for each specific ECG record, the computed threshold for detecting QRS occurrences.

The detection or adjustment of the most convenient scale factor is intrinsically related to the adaptation of the threshold parameter. Concerning a specific filtered signal version, the approach begins with an initial threshold value equal to 20% of the maximum amplitude within the filtered signal version. Then, if the number of the detected QRS occurrences is lower than half of the duration of the signal interval in seconds, the threshold value is decreased by 20%, and we search again for QRS fiducial points. Though unlikely, if the number of detected QRS fiducial points is higher than two times the duration of the signal interval in seconds, then the referred threshold is increased by 20%, and we search again for QRS fiducial points.

Once detected the fiducial points over a particular version of the preprocessed 10-s ECG excerpt, an evaluation metric is computed, based on the product between the standard deviation of the beat-to-beat intervals and the standard deviation of the peak amplitudes. The selected scale factor is that one associated with the filtering process, which allows for detection of fiducial points for which the referred evaluation metric has the lowest value for the test set.

We may observe that, according to the prevailing morphology and/or to the morphology variability, the scale factor adapts itself, such that most QRS points as possible are detected.

Concerning preprocessing stage, Ghaffari et al. (2009) and Martínez et al. (2004) proposed applying discrete wavelet transform (DWT) through filter banks, based on á trous algorithm, which works as a linear band-pass filtering. The authors generated five versions of the filtered signals corresponding to the scale factors: 2^1, 2^2, 2^3, 2^4, and 2^5. The difference between Mallat's algorithm and *algorithme à trous* is that Mallat's algorithm provides downsamplers after each filter to remove the redundancy of the signal representation. To keep the time-invariance and the temporal resolution at different scales, the *algorithme à trous* removes the decimation stages and interpolates the filter-impulse responses of the previous scale.

Both teams of authors used, as prototype wavelet $\Psi(t)$, a quadratic spline identified as the derivative of the convolution of four rectangular pulses, that is, the derivative of a low-pass function. Therefore the zero-crossings of the filtered signal versions correspond to the local maxima or minima of the smoothed signal at different scales, and the maximum absolute values of the wavelet transform are associated with maximum slopes in the original and smoothed signal. The transfer functions $G(z)$ and $H(z)$, respectively, are related to high-pass and low-pass filters and are obtained from the following equations:

$$H(e^{j\omega}) = e^{j\omega/2}\cos^3(\omega/2), \tag{4.7}$$

$$G(e^{j\omega}) = 4je^{j\omega/2}\sin(\omega/2), \tag{4.8}$$

which provide

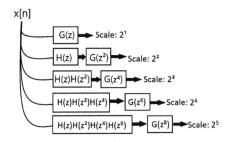

FIGURE 4.5

Filter bank implementation related to discrete wavelet transform based on á trous algorithm.

$$h[n] = \frac{1}{8}\delta[n+2] + 3\delta[n+1] + 3\delta[n] + \delta[n-1], \tag{4.9}$$

$$g[n] = 2\delta[n+1] - \delta[n]. \tag{4.10}$$

To implement DWT through *á trous* wavelet transform, high-pass and low-pass filters should be used according to the block diagram represented in Fig. 4.5.

According to Martínez et al. (2004), for frequency contents of up to 50 Hz, the *á trous* algorithm can be used in different sampling frequencies, which implies a certain independency of its results from sampling frequency, due to the fact that the main frequency contents of the ECG signal concentrate on the range less than 50 Hz. In this sense, Martínez et al. (2004) searches for QRS complex occurrences at scales from 2^1 to 2^4, through maximum modulus lines exceeding some thresholds.

Figs. 4.6(A), (B) illustrate the discrete wavelet transform of 6-s excerpts of ECG signals extracted from MIT–BIH Arrhythmia Database. As we can observe, the R-waves or the wave with the highest amplitude within the QRS complex are mapped as zero-crossings at each scaled version. Concerning QRS morphology, we remark that different morphologies provide different sequences of critical point signs around the zero-crossing, which is applied by Martínez et al. (2004) to automatically recognize QRS morphology patterns.

Also, it should be emphasized that the physiological and noise content determine which scales are more convenient to search QRS complex. For example, at record 200 (Fig. 4.6(A)), we may be able to identify QRS complexes at all the scaled versions, whereas for record 203 (Fig. 4.6(B)), only the last scaled version (scale 2^5) provides a non-confused observation.

As a nonlinear transformation, Ghaffari et al. (2009) applies over each scaled version a sliding window, aiming to compute a measure named area-curve length. So given a window with the length of L samples, sliding sample-to-sample on the signal excerpt related to a specific scaled version of the filtered signal, the kth segment can be obtained as

$$y_k = W_{2^\lambda}[k : k + L], \tag{4.11}$$

where y_k is a vector, including the samples k to $k + L$ of the filtered version related

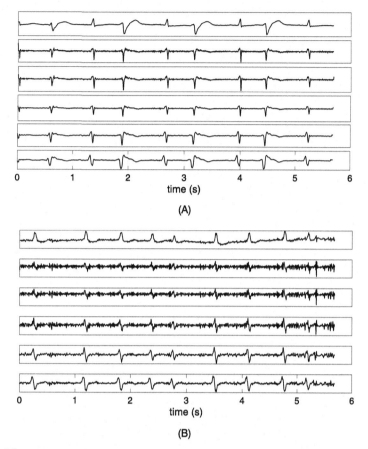

FIGURE 4.6

Scaled versions related to Discrete Wavelet Transforms of excerpts of ECG signals extracted from MIT–BIH Arrhythmia Database: Scales 2^1 to 2^5. (A) A 6-s excerpt of the raw signal and five scaled versions: record 200 (MIT–BIH Arrhythmia Database). (B) A 6-s excerpt of the raw signal and five scaled versions: record 203 (MIT–BIH Arrhythmia Database).

to scale 2^λ. Then, for each value of k, one defines the area under the absolute value of the time-series y_k and the curve length of y_k as

$$Area[k] = \int_{t_{0k}}^{t_{fk}} |y_k(t)|dt, \tag{4.12}$$

$$Curve[k] = \int_{t_{0k}}^{t_{fk}} \sqrt{1 + \dot{y}_k^2} = \sum_{t_{0k}+1}^{t_{fk}} \sqrt{1 + (y_k(n) - y_k(n-1))^2}, \tag{4.13}$$

where t_{0k} and t_{fk} are, respectively, the start and end points of the sequence y_k.

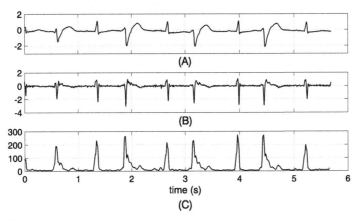

FIGURE 4.7

Process of linear and nonlinear transformation applied over a 6-s excerpt of ECG signal pertaining to Arrhythmia Database (record 200): (A) accessing raw signal, (B) filtering signal using DWT, scale 2^3, and (C) measuring area-curve length (ACL).

Thus the time-series measure named area-curve length (ACL) is defined as

$$ACL[k] = Area[k] \cdot Curve[k]. \qquad (4.14)$$

Concerning QRS detection and delineation, the authors apply DWT and obtain a filtered version of the signal, with scale factor equal to 2^3 or 2^4, but preferably 2^3. Then a window with the length of L around 40–50 ms is slid sample-to-sample on the signal, and the measure ACL is computed in each window.

Figs. 4.7(A)–(C) illustrate an example of applying DWT and computing the measure ACL over the record 200 from MIT–BIH Arrhythmia Database, which presents occurrence of a rhythm known as bigeminy.

According to Ghaffari et al. (2009), the most significant reason for the definition of the ACL measure is its ECG wave edges detection capability (delineation), or onset and offset locations. The proposed metric achieves its minimum value when both the value of amplitude of the filtered signal and the corresponding derivative in the window reach their minimum values. Thus a minimum value for the metric ACL indicates a minimum amplitude and minimum slope event.

Figs. 4.8(B), (C) illustrate another three examples of computing the metric ACL over 6 s excerpts of ECG signal, including conditions of low SNR and alternance of QRS morphologies.

Fig. 4.8 shows that high-frequency noise and QRS morphology alternance may impact and influence the resultant filtered signal related to the metric ACL, which can damage QRS-detection algorithm.

Arzeno et al. (2008) analyzed traditional first-derivative-based squaring functions and Hilbert transform-based methods for QRS detection and their modifications with improved detection thresholds, considering just single-lead ECG recordings. The authors emphasize that algorithms based on the differentiated ECG are computationally

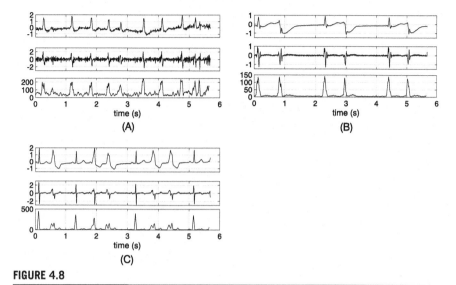

FIGURE 4.8

Linear (DWT) and nonlinear (metric ACL) filtering over ECG signal excerpts from MIT–BIH Arrhythmia Database containing different morphologies and SNR. (A) ECG 203, MIT–BIH Arrhythmia Database. (B) ECG 207, MIT–BIH Arrhythmia Database. (C) ECG 208, MIT–BIH Arrhythmia Database.

efficient for real-time analysis of large datasets, and that Hilbert transform-based algorithms and a modified version of the Hamilton–Tompkins algorithm (Hamilton and Tompkins, 1986) present an advantage for real-time applications by avoiding human intervention in threshold determination. The high accuracy of the Hilbert transform-based methods for emphasizing and detecting QRS complex is ascribable, as already explained, to its inherently uniform magnitude spectrum.

According to Arzeno et al. (2008), among several tested approaches, detection errors occurred mainly in beats with decreased signal slope, such as wide arrhythmic beats or attenuated beats, such that differences in detection will point to abnormalities in the signal that can be further analyzed. Before obtaining the first derivative of the QRS complex, the ECG was bandpass-filtered with a Kaiser window filter, with pass-band 8 to 20 Hz to remove baseline wander and high-frequency noise.

The ideal bandpass filter may be described in the frequency domain as

$$
H_d(e^{j\omega}) = \begin{cases} e^{j\omega} \begin{cases} -\omega_{c2} - \omega_{c1} < \omega < -\omega_{c2} + \omega_{c1}, \\ \omega_{c2} - \omega_{c1} < \omega < \omega_{c2} + \omega_{c1}, \end{cases} \\ 0 \text{ otherwise.} \end{cases} \tag{4.15}
$$

Transforming $H_d(e^{(j\omega)})$ to the time-domain and considering N as the filter order and $M = N/2$ yield

$$
h_d[n] = \begin{cases} \frac{\sin(\omega_{c2}(n-M))}{\pi(n-M)} - \frac{\sin(\omega_{c1}(n-M))}{\pi(n-M)}, n \neq M, \\ \frac{\omega_{c2} - \omega_{c1}}{\pi}, n = M. \end{cases} \tag{4.16}
$$

The Kaiser window is defined as

$$w[n] = \begin{cases} \frac{I_0[\beta(1-[(n-\alpha)/\alpha]^2)^{1/2}]}{I_0(\beta)}, & 0 \le n \le M, \\ 0 \text{ otherwise,} \end{cases} \tag{4.17}$$

where $\alpha = M/2$, and $I_0(\cdot)$ represents the zeroth-order modified Bessel function of the first kind (Oppenheim, 1999).

In the first methodology, Arzeno et al. (2008) applied in a serial sequence bandpass filtering, with Kaiser window, the first-derivative (implemented as a center differentiation stage), created zero-crossings in the location of the R-peaks and Hilbert transform in order to rectify the phase providing outstanding peaks in the location of R-peaks.

In a second methodology, the authors proposed, after bandpass filtering, a five-point derivative formula, given as

$$v_r[n] = \frac{1}{8}(2x[n] + x[n-1] - x[n-3] - 2x[n-4]), \tag{4.18}$$

which prevents high-frequency noise amplification, a squaring of the differentiated signal, and then time-averaging by taking the mean of the previous 32 samples. At a subsequent stage, peaks may be found by comparing the time-averaged signal to a primary threshold.

Although squaring function provides additional attenuation of other ECG features, leaving the QRS complexes as outstanding positive peaks in the signal, regardless of their polarity and morphology, its major disadvantage is that normal QRS peaks with small amplitudes and wide complexes (with decreased slope) are reduced in the output of the transform.

Figs. 4.9(A), (B) illustrate the two filtering-related methodologies, both applying bandpass filtering and squaring function. It must be emphasized the noise attenuation capability of the second methodology outweighs the first methodology, but in comparison, almost neglect the amplitude of wide arrhythmic beats.

In the third methodology (Arzeno et al., 2008) employs the squaring function as the main transformation, not performing the time-averaging, and in the fourth methodology the authors implement a second differentiation stage, as a forward derivative, also as the main transformation, which creates a one-sample delay and preserves most of the energy of the high frequencies corresponding to the QRS complex. The second differentiation stage rectifies the signal by its odd-phase property, resulting in a large local minimum at each QRS fiducial point.

Fig. 4.10 illustrates a practical case of those two described methods. It should be emphasized the discrepancy between the filtering responses, considering normal and arrhythmic beats, and also that the odd-phase and high-pass characteristics of the second-derivative results in nonuniform filtering of the low and high frequency components of the differentiated ECG.

As already explained in the previous chapter, the empirical mode decomposition (EMD) is commonly applied for ECG denoising and, therefore, also for QRS enhancing. Several approaches apply empirical mode decomposition for that purpose, as Pal

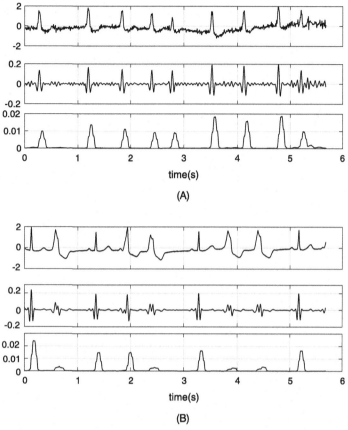

FIGURE 4.9

Linear and nonlinear filtering over ECG signal excerpts from MIT–BIH Arrhythmia Database containing different morphologies and SNR: each graph presents raw signal, first preprocessing methodology (band-pass filtering, first-derivative and Hilbert transform) and second methodology (band-pass filtering, five-point derivative, squaring, and time-averaging). (A) ECG 203, MIT–BIH Arrhythmia Database. (B) ECG 208, MIT–BIH Arrhythmia Database.

and Mitra (2010) and Slimane and Naït-Ali (2010). Remember that, the EMD of an original ECG signal $x[n]$ is given as

$$x[n] = \sum_{k=1}^{N} c_k[n] + r_N[n], \tag{4.19}$$

where each $c_k[n]$ is an intrinsic mode function (IMF), and $r_N[n]$ is a residue.

As is well known, the EMD relies on a fully data-driven mechanism, and it does not require any a priori knowledge about the signal content, or basis functions, and even sampling frequency. However, the full number of intrinsic mode functions de-

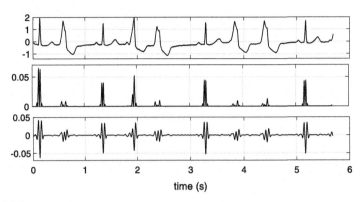

FIGURE 4.10

Linear and nonlinear filtering over an ECG signal excerpt from MIT–BIH Arrhythmia Database (record 208): raw signal, third preprocessing methodology (band-pass filtering, first-derivative and squaring) and second methodology (band-pass filtering, first derivative, and second differentiation stage).

pends on physiological and noise content. In this sense, Slimane and Naït-Ali (2010), after applying baseline wander cancellation, it takes the first three IMFs: $c_1[n]$, $c_2[n]$, and $c_3[n]$, and for each component, a nonlinear transform is computed as follows:

$$y_i[n] = \begin{cases} |c_i[n]c_i[n-1]c_i[n-2]|, & c_i[n]c_i[n-1]c_i[n-2] > 0, \\ 0 \text{ otherwise,} \end{cases}$$ (4.20)

where the index i may assume the values $i = 1, 2, 3$.

Then each resultant signal is presented to a moving average process, in order to produce a signal that includes information about QRS slope, width, and morphology. By using a number of samples M, a moving window integrator is computed for each $y_i[n]$ as

$$S_i[n] = \frac{1}{M}(y_i[n-(M-1)] + y_i[n-(M-2)] + ... + y_i[n]).$$ (4.21)

The authors suggest that M should be related to the duration of the widest possible QRS.

Finally, a resultant output is computed as the summation of each obtained $S_i[n]$, given as

$$S_t[n] = \sum_{i=1}^{3} S_i[n].$$ (4.22)

Figs. 4.11(A)–(D) illustrate the processed resultant signal, considering the referred to approach, for signal excerpts from the records 200, 203, 207, and 208, all from MIT–BIH Arrhythmia Database. Each processed signal is related to the summa-

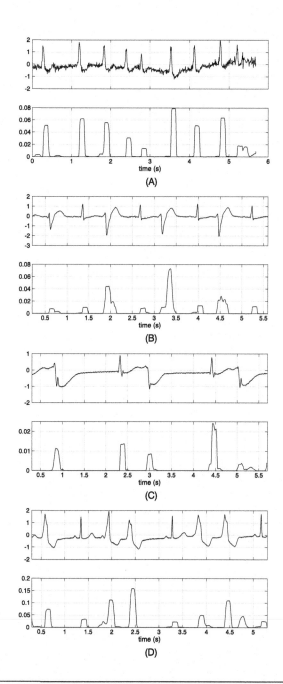

FIGURE 4.11

QRS complex enhancing based on empirical mode decomposition, linear and non-linear filtering of each intrinsic mode function. (A) ECG 200, MIT–BIH Arrhythmia Database. (B) ECG 203, MIT–BIH Arrhythmia Database. (C) ECG 207, MIT-BIH Arrhythmia Database. (D) ECG 208, MIT-BIH Arrhythmia Database.

tion of the time-averaged and processed intrinsic mode functions. Each figure shows the temporal mapping between original QRS complexes and filtered beats.

We observe, by way of each resultant filtered signal, that each outstanding peak provides information concerning where a significant energy increase occurred. Although the amplitude of the enhanced peaks varies abruptly, according to the amplitude and morphology of the original QRS complexes, which may require an adaptive threshold computing at decision stage, different patterns of amplitude, duration, and area of each energy wave may be associated, or match corresponding different patterns of QRS morphology.

4.3 DECISION STAGE

The decision stage receives as input information the filtered and enhanced signal, which results from the preprocessing stage. Decision rules are applied to the preprocessed signal in order to decide whether or not a QRS complex has occurred. Once the preprocessing stage is prone to eventual failure, the decision stage should avoid, as much as possible, the occurrences of false-positive and false-negative detections. Generally, the algorithms at decision stage are based on an adaptive threshold, that is, each sample of the filtered signal has its amplitude compared with a reference amplitude level, which should incorporate adaptivity, since QRS amplitude and morphology may change drastically during certain signal excerpts, or even among consecutive beats. As detailed and explained by Sörnmo and Laguna (2005), a fixed threshold inevitably implies accepting several false detections. Furthermore, although adaptive threshold approaches concerns amplitude-based rules, we commonly find in the literature additional rules (some of them, empirical ones), related to other signal properties, such as beat-to-beat intervals, and the duration of a waveform.

Now, we will describe several approaches from the literature for threshold-based decision stages, which are intrinsically related to specific methodologies employed for QRS enhancing, and compare the overall performances over well-known standard ECG datasets. We will also present and illustrate some case studies for comparing and testing a subset of referenced approaches, which may involve different levels of noise incidence, QRS morphology diversity, and single or multiple ECG channels.

Before introducing the description and the performance comparison for different approaches, it is important to comment about applied metrics for performance evaluation. These parameters may be determined and evaluated before a QRS detector can be implemented in a clinical setting. According to Sörnmo and Laguna (2005), QRS detector performance is commonly measured in terms of:

- P_D, the probability of a true beat being detected, and
- P_F, the probability of a false beat being detected.

In practice, these probabilities are estimated on the basis of computing simulation results over multiple ECG databases with a large variety of QRS morphologies and noise types. In this case, the estimation is based on ratios involving the number of correctly detected QRS complexes, also known as true positives, N_D or T_P, the number of false alarms or false positives, N_F or F_P, and the number of missed beats or false negatives, N_M or F_N.

In this sense, the probability of detection is estimated from Sörnmo and Laguna (2005) as

$$\hat{P}_D = \frac{N_D}{N_D + N_M}. \qquad (4.23)$$

This probability estimation is also known as sensitivity (Se) (Kohler et al., 2002). On the other hand, the probability of false detection can be estimated from Sörnmo and Laguna (2005)

$$\hat{P}_F = \frac{N_F}{N_D + N_F}. \qquad (4.24)$$

The complement of this probability is named positive predictivity ($P+$, that is, the probability of a given detection be a true detection), computed as

$$P+ = \frac{N_D}{N_D + N_F}. \qquad (4.25)$$

We should emphasize that the computing of the quantities N_D, N_F, and N_M requires manual annotation of the beats (beat locations) for a given database, which is a typically laborious process, involving one or several specialist ECG readers. By manual annotation process, every QRS complex, from a record and/or from a dataset, is assigned to its correct occurrence time T_i (Sörnmo and Laguna, 2005). A beat is considered to have been detected when the difference between the estimated/computed occurrence time, $\hat{T}_j$ and the annotation time T_i is within a certain tolerance window, defined by ΔT, that is,

$$|\hat{T}_j - T_i| \le \Delta T. \qquad (4.26)$$

A false detection, or a false-positive, is identified when $\hat{T}_j$ is located at a time distance larger than ΔT from any T_i, and a beat is considered to have been missed, a false-negative occurrence, when no detection occurs closer than ΔT to T_i.

According to Sörnmo and Laguna (2005), an adaptive threshold for QRS detection, is, in general, interval-dependent, that is, it is updated once for each new detection and is held fixed during the subsequent R–R interval, until the current threshold value is exceeded, and a new QRS is detected. In this sense, Madeiro et al. (2012) employs an adaptive threshold technique for detecting QRS fiducial points over the ECG raw signal, that is, non-preprocessed ECG signal. For applying this approach, the methodology departs from data obtained within a 10-s training interval. The data corresponds to the average $\bar{m}(R - R)$ and the standard deviation $\sigma(R - R)$ of the intervals between already detected QRS fiducial points. The expression for adaptive threshold is given as

$$th[k] = \frac{\beta_1 Re[k] + \beta_2 R[k-1]}{\beta_1 + \beta_2} \alpha, \qquad (4.27)$$

where $th[k]$ is the threshold value computed to detect the kth QRS fiducial point, β_1 and β_2 are weight factors, $Re[k]$ is an estimation for the magnitude of the amplitude of the kth beat, based on the previous value of $th[k]$; $R[k-1]$ is the magnitude of the amplitude of the $(k-1)$th beat, and α is a percentage factor.

The adaptive threshold is initialized as $th[k_0]$ just after the training stage according to the expression

$$th[k_0] = \frac{|\bar{R}| + |R[k_0 - 1]|}{2} \alpha, \tag{4.28}$$

where $k_0 - 1$ is the number of QRS fiducial points detected in the training stage, $|\bar{R}|$ is the average of the magnitude of amplitudes of QRS fiducial points detected in the training stage, and $|R[k_0 - 1]|$ is the magnitude of the amplitude of the last QRS fiducial point detected in the training stage.

The identification of a beat is accomplished by the detection of the QRS fiducial point. According to Madeiro et al. (2012), the detection of a QRS fiducial point is implemented by first identifying the first and the last samples, which exceed in module the threshold level. Then the sample between them having the highest amplitude in module is taken as the QRS fiducial point.

Therefore when a peak is detected exceeding the threshold $th[k]$ as a candidate for the kth beat, the algorithm evaluates its reliability through a statistic metric M_T given as

$$M_T = \frac{I[k] - \bar{m}(R - R)}{\sigma(R - R)}, \tag{4.29}$$

where $I[k]$ refers to the interval between the $(k-1)$th QRS fiducial point and the candidate for the kth beat.

Two signal-opposed limits are set as tolerance criteria for M_T: λ_1 and λ_2. If $\lambda_1 \leq M_T \leq \lambda_2$, then the candidate for the kth beat is accepted as the kth QRS fiducial point.

After this, the average and standard deviation of intervals between beats are updated, and a new value for the adaptive threshold is computed for the subsequent R–R interval. Otherwise, if $M_T \leq \lambda_1$ or $M_T \geq \lambda_2$, then two filtering processes are performed, one for each situation, based on wavelet and Hilbert transforms and first derivative. At the first situation, when $M_T \leq \lambda_1$, then the algorithm defines a time-window containing the last detected QRS fiducial point and the candidate for the kth beat. Then the same filtering techniques, which were applied at the training stage are again implemented over the referred to window. Now, the goal is to verify if the detected peak appears in the filtered version of the signal excerpt exceeding a specific threshold. Once we verify the detected peak, it is mapped within the filtered window, then the candidate for the kth beat is accepted as the kth QRS fiducial point. After that, the average and standard deviation of intervals between beats are updated, and a new value for the adaptive threshold is computed for the subsequent R–R interval. Otherwise, if we do not verify any peak exceeding the computed threshold within the filtered signal, the detected peak is discarded, and the search for new beats is commenced.

In the second situation, when $M_T \geq \lambda_2$, then the algorithm defines a time window between the $(k-1)$th detected beat and the peak, which is candidate for the kth beat. Analogously, the same filtering techniques, which were applied at the training stage are again implemented over the referred to window. Here, the goal is to verify if an eventual beat, which was not detected by the adaptive threshold within the raw signal, appears in its filtered version exceeding a specific threshold. Once we identify one or more peaks in the filtered signal, they are respectively mapped within the raw signal, and accepted as the kth and subsequent QRS fiducial points. Because we lose time resolution through filtered signal versions, as scale factor of the wavelet transform increases, the mapping searches for critical points within the raw signal, which are located at a maximum temporal distance of 120 ms from the reference instant t_p, where we detect peak exceeding a threshold within the filtered signal. Then we associate the critical point with the maximum amplitude (in module) to the next detected QRS fiducial point.

We can enumerate some strengths related to the algorithm for QRS detection developed by Madeiro et al. (2012): algorithm running speed, robustness in face of noise interference, QRS morphology changes, and cardiac rhythm changes. In fact, for records with regular rhythms, the preprocessing rate plummets to minimum levels, that is, the algorithm rarely requires filtering routines. For example, considering the record 100 from MIT–BIH Arrhythmia Database, which presents a length of 30 minutes, the referred to algorithm requires just 1.53 seconds to detect all the QRS complexes with a 100%-sensitivity rate, and a 100% positive predictivity rate by applying the described algorithm over the first channel. The computed preprocessing rate is equal to 13.79%. In order to operate the algorithms through simulation tests with digitized ECG signals, freely available in several databases from the MIT–BIH, we use a 2.70 GHz Intel(R) Core(TM) i7-7500U CPU, 8 GB of RAM and the Matlab® R2017b.

Obviously, the more intense the amplitude or morphological variability of the QRS complex, the more recurrent the filtering procedures, and each specific ECG record will demand a variable processing time and a variable preprocessing rate. For example, if we apply the algorithm individually to the two available ECG channels of the record 100 of the MIT–BIH Arrhythmia Database, we find for the second channel a preprocessing rate equal to 16.34% and a required processing time of 1.63 seconds.

Figs. 4.12(A)–(C) illustrate some results of well-succeeded QRS detections, by marking R-waves with circles, using the algorithm developed by Madeiro et al. (2012), concerning detections implemented individually over each available channel and the derived R–R interval time-series. Here it is worth emphasizing that the vast majority of algorithms for QRS detection reported in the literature only uses a single channel of the available ECG signals as input information, this being, in general, the first channel.

An important issue related to QRS detection is the identification of a stable and accurate reference for QRS fiducial point. According to Sörnmo and Laguna (2005),

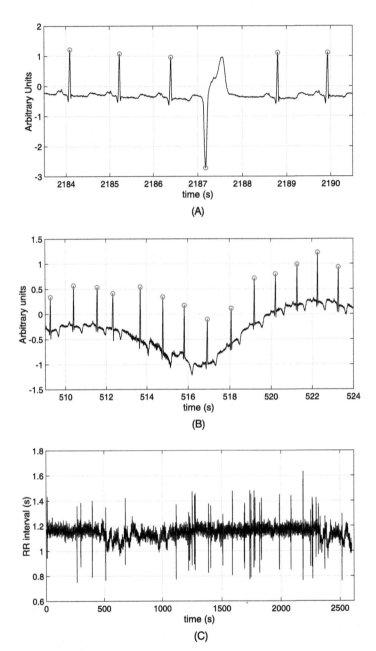

FIGURE 4.12

Results for the QRS detection algorithm developed by Madeiro et al. (2012) illustrating beats detected over each available channel and the computed R–R intervals time-series. (A) QRS detection over record 100/mitdb: first channel. (B) QRS detection over record 100/mitdb: second channel. (C) Time-series of R–R intervals detected for record 100/mitdb.

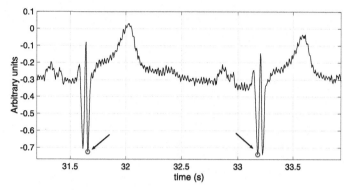

FIGURE 4.13

Example of low temporal resolution issue concerning QRS detection by applying the algorithm developed by Madeiro et al. (2012) within the record 114/mitdb: QRS complex markings are either taken as the second critical point, or as the fourth critical point.

as already stated, the majority of QRS detectors are designed to detect heart beats, though rarely producing occurrence times with high temporal resolution. Depending on the QRS morphology, the critical point within the QRS complex with the highest amplitude in module does not behave as a stable reference. In these situations, it may be necessary to use an algorithm that performs time alignment of the detected beats.

A classical example of low temporal resolution issue concerning QRS detection is illustrated by applying the algorithm developed by Madeiro et al. (2012) within the record 114/mitdb. As Fig. 4.13 illustrates, the automatic markings related to QRS fiducial points are either taken as the second critical point (local minimum) within the QRS complex, or taken as the fourth critical point (local minimum).

The approach developed by Martínez et al. (2004), as already detailed, obtains filtered versions of the ECG signal concerning Wavelet transform related to scales 2^1 to 2^4. The algorithm searches across the scales for "maximum modulus" exceeding some thresholds at each respective scale: ε^1_{QRS}, ε^2_{QRS}, ε^3_{QRS}, and ε^4_{QRS}. The algorithm rejects isolated and redundant maximum lines, and the zero crossing of the wavelet transform at scale 2^1, between a positive maximum–negative minimum pair, which should be identified among the four referred to scaled versions, is marked as a QRS. According to the authors, some protection measures are taken, like a refractory period or a search back with lowered thresholds, if a significant time has elapsed without detecting any QRS. The thresholds are not updated for each beat, but for each excerpt/interval of 2^{16} samples.

As the algorithm takes as QRS fiducial point reference the zero-crossings instants, which are identified within the first scaled version, the approach minimizes the issue of accurate time resolution. In this sense, Fig. 4.14 illustrates, at the upper graph, a signal excerpt from the record 101/mitdb, containing a complete ECG cycle and its R-wave peak identified with a circle. In the second graph, we can observe the four wavelet transform scaled versions of the ECG cycle and the sample related to the

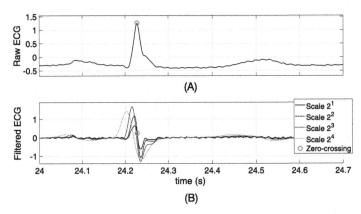

FIGURE 4.14

Signal excerpt from the record 101/mitdb, containing: (A) a complete ECG cycle and its R-wave peak identified with a circle, and (B) the four wavelet transform scaled versions of the ECG cycle and the sample related to the R-wave peak, recognized as the sample immediately before the zero-crossing within the first scaled version.

R-wave peak, recognized as the sample immediately before the zero-crossing within the first scaled version.

As we can observe, in the approach developed by Martínez et al. (2004), isolated maximum or minimum lines are generally related to artifacts. In this sense, QRS complexes should be mapped with at least two peaks of opposed signs (positive maximum–negative minimum pair) within each scaled version. Therefore Fig. 4.15(A)–(C) illustrates, in the upper graph, a sequence with four consecutive detected beats and the presence of artifacts, which abruptly change baseline levels, and, in the lower graphs, the wavelet transform concerning scaled versions 2^1 and 2^4, with their corresponding threshold levels. Applying the decision algorithm, which associates significant positive maximum–negative minimum pairs (in each scaled version), with maximum slopes within the QRS complexes and zero-crossings between them, with QRS fiducial points, we accurately detect instants of QRS complexes.

It should be emphasized that, according to the approach referred to, for each specific QRS morphology, the number and sequence of positive/negative maximum/minimum critical points within the wavelet transform will change, which requires routines within the algorithm to keep stable and accurate detections for QRS fiducial points.

As already detailed previously, we can enumerate the following possible QRS morphologies: qR, qRs, Rs, R, RS, rSR', rR', qrSr', RSr', rR's, rS, rSr', Qr, QS, QR, qrS, qS, rsR's', QRs, and Qrs. The occurrence of each specific morphology may be linked to a specific lead used to acquire the ECG signal or even to a given arrhythmia or cardiac disease. Thus it is also of vital importance to recognize the predominant QRS morphology or the corresponding changes within a given ECG record.

As we can observe in Fig. 4.16, the record sel117/qtdb presents a predominant RS-morphology for QRS. Its wavelet transform appears as a sequence of three criti-

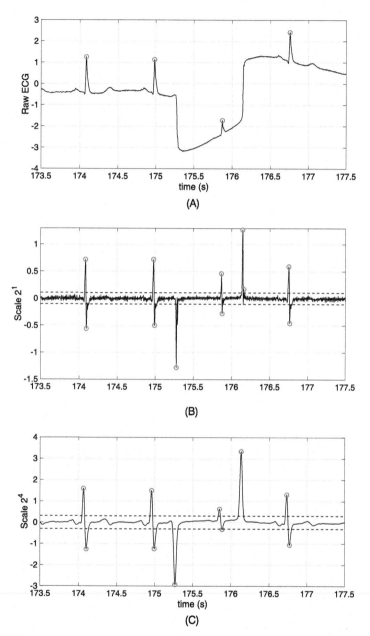

FIGURE 4.15

Decision process which associates positive maximum–negative minimum pairs (in each wavelet transform scaled version) with maximum slopes within the QRS complexes and zero-crossings between the critical points with QRS fiducial points: (A) raw ECG signal, (B) scaled version 2^1, and (C) scaled version 2^4.

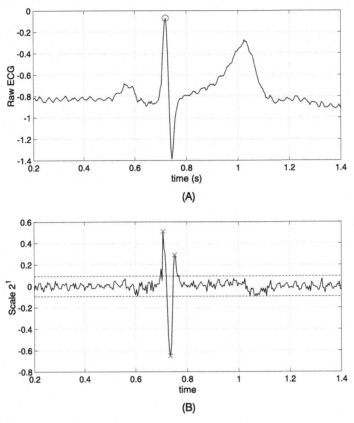

FIGURE 4.16

Identification of RS-morphology for a given QRS complex within the record sel117/qtdb: (A) raw ECG signal, (B) scaled version 2^1 presenting three critical points.

cal points: positive–negative–positive sequence. The R-wave peak is mapped as the zero-crossing between the critical points with maximum energy (the two with highest amplitude).

As a second example (see Figs. 4.17(A), (B)), the record sel114/qtdb presents a predominant rsR's'-morphology for QRS. Its wavelet transform appears as a sequence of five critical points: positive–negative–positive–negative–positive sequence. The R-wave peak is mapped as the zero-crossing between the third and the fourth critical points (the two with highest amplitude, excluding the latter).

Some features and characteristics from the approach developed by Martínez et al. (2004) make it appropriate for offline ECG analysis, such as adaptation of the threshold value for each excerpt of 2^{16} samples.

Concerning the approach proposed by Ghaffari et al. (2009), after applying DWT and defining a window with the length of L around 40–50 ms, which is slid sample-

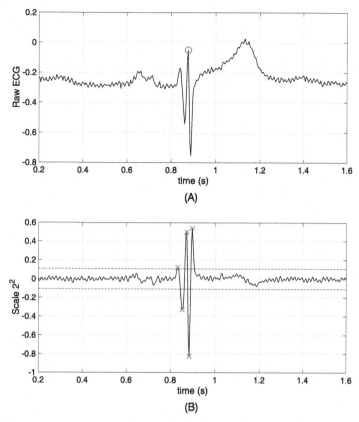

FIGURE 4.17

Identification of rsR's'-morphology for a given QRS complex within the record sel114/qtdb: (A) raw ECG signal, (B) scaled version 2^2 presenting five critical points.

to-sample on the filtered signal for computing the metric ACL (Eqs. (4.11) to (4.14)), the signal ACL is divided into segments of 400- to 500-ms samples. Then for each window segment within the signal ACL, one computes the mean amplitude value μ_i, standard deviation σ_i of the amplitude values, and a variable threshold given as $\tau_i = \mu_i + \alpha\sigma_i$, where α, $1 \leq \alpha \leq 4$, is a comparison coefficient. This variable threshold is applied for identifying each R-wave peak location, which is associated with the position maximum values (critical points) exceeding instantaneous/current values of the variable threshold.

Thus the threshold parameter here is not updated at each QRS detection, and it does not depend on previous values, but it is a function of the amplitude values related to each window segment. The authors do not provide a single specific value for α, arguing that the proper interval for assigning values to α is obtained after numerous simulations, that is, empirically.

As some examples of QRS detection concerning different QRS morphologies and also physiological noise, Figs. 4.18(A)–(C) present application of the approach proposed by Ghaffari et al. (2009) over records 100, 108, and 200, from MIT–BIH Arrhythmia Database. For these computing simulation results, one considers $\alpha = 2$, and segments of 400 ms. Observing the illustrated results, we verify that segment-based threshold signal has a significant sensibility for recognizing amplitude variations due to QRS complex. Sometimes a unique threshold value encompasses an entire QRS complex, other times only a fraction of a beat segment. But even when a single QRS is comprised of two threshold levels, the corresponding values potentially avoid false-positive detections. Despite the authors do not comment about protection measures, we should consider refractory periods or search back with lowered thresholds, if significant time has elapsed without QRS detections.

4.4 EVALUATING AND COMPARING PERFORMANCES OF QRS DETECTORS

It is well-known that applying QRS detection algorithms in medical devices requires the evaluation of detection performance. Several standard and annotated ECG databases are available for the evaluation of software QRS detection algorithms, providing means for comparing and reproducing results. One of the great advantages of using standard databases is that they contain a large number of selected signals representative of a wide variety of physiologic conditions, as well as signals, which are rarely observed but clinically important (Kohler et al., 2002).

Most frequently the MIT–BIH Arrhythmia Database is used Goldberger et al. (2000). It pertains to a major database named MIT–BIG database, provided by MIT and Boston's Beth Israel Hospital, which contains ten databases for various test purposes: the Arrhythmia Database, the Noise Stress Test Database, the Ventricular Tachyarrhythmia Database, the ST Change Database, the Malignant Ventricular Arrhythmia Database, the Atrial/Fibrillation/Flutter Database, the ECG Compression Test Database, the Supraventricular Arrhythmia Database, the Long-Term Database, and the Normal Sinus Rhythm Database (Kohler et al., 2002; Goldberger et al., 2000). The MIT–BIH Arrhythmia Database contains 48 half-hour records of ECG signals, with a sampling rate of 360 Hz and 11-bit resolution over a 10-mV range. Altogether there are 116137 QRS complexes, and each record file provides two ECG channels (leads). As has been indicated, the universe of QRS complexes presents several difficulty levels for automatic detections, including since clear R-peaks and few artifacts until QRS complexes with abnormal shapes, give rise to morphology alternance, noise, and artifacts. Twenty-five records with less common arrhythmias were selected from over 4000 24-hour ambulatory ECG records, and the rest were chosen randomly (Kohler et al., 2002).

Other frequently applied database for evaluation purposes is the QT Database, available on PhysioBank website (Laguna et al., 1997). It was designed for evaluation of algorithms that detect waveform edges, and also, for QRS detection. Altogether,

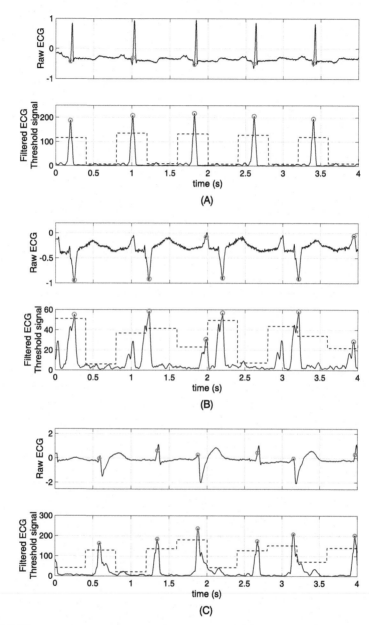

FIGURE 4.18

Decision process which associates maximum positive peaks exceeding instantaneous values of a variable threshold with QRS complex occurrences: (A) raw ECG, filtered ECG and threshold signal for record 100/mitdb, (B) raw ECG, filtered ECG and threshold signal for record 108/mitdb, and (C) raw ECG, filtered ECG and threshold signal for record 200/mitdb.

Table 4.1 Performance comparison of several QRS detection algorithms: application to MITDB

Detection algorithm	Annotations	TP	FP	FN	Error (%)	Se (%)	P+ (%)
Madeiro et al. (2012)	109,495	108,568	856	928	1.69	99.15	99.18
Madeiro et al. (2007)	109,494	107,808	1073	1686	2.57	98.47	98.96
Ghaffari et al. (2009)	109,428	109,327	129	101	0.21	99.91	99.88
Martínez et al. (2004)	109,428	109,208	153	220	0.34	99.80	99.86
Slimane and Naït-Ali (2010)	110,050	109,876	84	174	0.23	99.84	99.92
Di Marco and Chiari (2011)	109,010	108,758	148	252	0.37	99.77	99.86
Martínez et al. (2010)	109,428	109,111	35	317	0.32	99.71	99.97

105 records with a wide variety of QRS and ST–T morphologies were selected from other databases, including European ST–T Database and MIT–BIH Arrhythmia Database. Each record has a length of 15 minutes and 250-Hz sampling frequency.

In order to make a comparison study, we rank a subset of the most current reported results considering the algorithms tested against MIT–BIH Arrhythmia Database, as synthesized within Table 4.1. From the listed results, we note that the sensitivity and positive predictivity rates (the main metrics) do not show great differences, and all the algorithms present the afore-mentioned predictivity rates around or above 99%. We may say that there are many methods, which provide almost negligible error rates. Thus considering the current status of QRS detection algorithms, the search for higher accuracy in locating QRS complexes is a research demand already sufficiently addressed.

However, current ECG processing methods (not only QRS detection, but also ECG delineation and beat classification), some of them based on machine-learning approaches, require sufficient computational resources. Are these methods fast and feasible in real-time applications? Are these methods applicable for smartphone environment and other battery-driven devices?

According to Elgendi (2013), there is a need to develop numerically efficient algorithms to accommodate the new trend towards battery-driven ECG devices and to analyze long-term recorded signals in a time-efficient manner. In this sense, despite the very high performance of the most current approaches for QRS detection, it is well-known that they offer a high computing load, comprising several preprocessing layers and a decision stage, which may require a previously stored signal window (offline analysis). Thus an important demand for research, not completely addressed nowadays, is to increase the efficiency of the algorithms, adapting them for real-time applications and battery-driven devices, for which there will always be a power-consumption limitation in processing ECG signals. The problem is to solve a trade-off between a significantly high performance and a fast and efficient algorithm, suitable for real-time application within mobile and battery-driven devices.

Furthermore, considering that QRS detection is the first task (first processing layer) required for automatic recognition of cardiac disease patterns, some errors related to false-positive and false-negative QRS detections are propagated to the following stages and have a significant impact in the final arrhythmia classification/recognition. According to Luz et al. (2016), a large set of beat classification approaches utilized databases in which the events related to heartbeat detection/segmentation are identified and previously labeled, reducing the segmentation stage to a simple search of a labeled event in the database. Thus the results reported by several works, which focus on arrhythmia detection and classification, disregard the impact of the QRS detection/segmentation step. Therefore another active demand is to research the impact of different QRS detection/segmentation algorithms on feature extraction and automatic arrhythmia classification methods.

4.5 CONCLUSIONS

In this chapter, we performed a literature review concerning the classical structure for QRS detection, including the wide diversity of "noise" from a QRS-detection point of view. We detailed the preprocessing stage, presenting several filtering techniques, including linear and nonlinear transformation, concerned in emphasizing QRS complex and attenuating physiological noise and interference. Then we looked at the issue of decision stage, that is, how to identify the QRS occurrence after we had applied preprocessing. We showed that many approaches suffer from problems related to loss time-resolution, that is, they do not provide a stable reference for QRS complexes. Furthermore, it is well-known that adaptive threshold is a commonly applied technique at decision stages, and its value may be adapted beat-after-beat or at each specific interval length. Finally, we verified that, despite the very high performance of the most current approaches for QRS detection, an important demand for research, not completely addressed nowadays, is to increase the efficiency of the algorithms, adapting them for real-time applications and battery-driven devices, for which there will always be a power-consumption limitation in processing ECG signals.

REFERENCES

Arzeno, N.M., Deng, Z.D., Poon, C.S., 2008. Analysis of first-derivative based QRS detection algorithms. IEEE Transactions on Biomedical Engineering 55 (2), 478–484.

Di Marco, L.Y., Chiari, L., 2011. A wavelet-based ECG delineation algorithm for 32-bit integer online processing. BioMedical Engineering OnLine 10 (1), 23.

Elgendi, M., 2013. Fast QRS detection with an optimized knowledge-based method: evaluation on 11 standard ECG databases. PLoS ONE e73, 557.

Ghaffari, A., Homaeinezhad, M., Akraminia, M., Atarod, M., Daevaeiha, M., 2009. A robust wavelet-based multi-lead electrocardiogram delineation algorithm. Medical Engineering & Physics 31 (10), 1219–1227.

Goldberger, A.L., Amaral, L.A., Glass, L., Hausdorff, J.M., Ivanov, P.C., Mark, R.G., Mietus, J.E., Moody, G.B., Peng, C.K., Stanley, H.E., 2000. PhysioBank, PhysioToolkit, and PhysioNet. Circulation 101 (23), e215–e220.

Hahn, S.L., 1996. Hilbert Transforms in Signal Processing, vol. 2. Artech House, Boston.

Hamilton, P.S., Tompkins, W.J., 1986. Quantitative investigation of QRS detection rules using the mit/bih arrhythmia database. IEEE Transactions on Biomedical Engineering 12, 1157–1165.

Kohler, B.U., Hennig, C., Orglmeister, R., 2002. The principles of software QRS detection. IEEE Engineering in Medicine and Biology Magazine 21 (1), 42–57.

Laguna, P., Mark, R.G., Goldberg, A., Moody, G.B., 1997. A database for evaluation of algorithms for measurement of QT and other waveform intervals in the ECG. In: Computers in Cardiology 1997. IEEE, pp. 673–676.

Lugovaya, T.S., 2005. Biometric Human Identification Based on ECG. PhysioNet.

Luz, E.J.d.S., Schwartz, W.R., Cámara-Chávez, G., Menotti, D., 2016. ECG-based heartbeat classification for arrhythmia detection: a survey. Computer Methods and Programs in Biomedicine 127, 144–164.

Madeiro, J.P., Cortez, P.C., Oliveira, F.I., Siqueira, R.S., 2007. A new approach to QRS segmentation based on wavelet bases and adaptive threshold technique. Medical Engineering & Physics 29 (1), 26–37.

Madeiro, J.P., Cortez, P.C., Marques, J.A., Seisdedos, C.R., Sobrinho, C.R., 2012. An innovative approach of QRS segmentation based on first-derivative, Hilbert and Wavelet Transforms. Medical Engineering & Physics 34 (9), 1236–1246.

Martínez, J.P., Almeida, R., Olmos, S., Rocha, A.P., Laguna, P., 2004. A wavelet-based ECG delineator: evaluation on standard databases. IEEE Transactions on Biomedical Engineering 51 (4), 570–581.

Martínez, A., Alcaraz, R., Rieta, J.J., 2010. Application of the phasor transform for automatic delineation of single-lead ECG fiducial points. Physiological Measurement 31 (11), 1467.

Oppenheim, A.V., 1999. Discrete-Time Signal Processing. Pearson Education, India.

Pal, S., Mitra, M., 2010. QRS complex detection using empirical mode decomposition based windowing technique. In: Signal Processing and Communications (SPCOM), 2010 International Conference on. IEEE, pp. 1–5.

Slimane, Z.E.H., Naït-Ali, A., 2010. QRS complex detection using empirical mode decomposition. Digital Signal Processing 20 (4), 1221–1228.

Sörnmo, L., Laguna, P., 2005. Bioelectrical Signal Processing in Cardiac and Neurological Applications, vol. 8. Academic Press.

Delineation of QRS Complex: Challenges for the Development of Widely Applicable Algorithms

João Paulo do Vale Madeiro*, José Maria da Silva Monteiro Filho†,
Priscila Rocha Ferreira Rodrigues†

*Institute for Engineering and Sustainable Development (IEDS), University for the International Integration of the Afro-Brazilian Lusophony – UNILAB, Redenção, Ceará, Brazil
†Department of Computing Science, Federal University of Ceara, Fortaleza, Ceará, Brazil

5.1 BASIC CONCEPTS FOR WAVE DELINEATION

According to Sörnmo and Laguna (2005), the classical definition of a wave edge (or boundary) is the instant at which the wave crosses a certain amplitude threshold level (increasing or decreasing). In practice, this definition is not applied due to baseline wandering and also due to the fact that near the wave boundaries, the noise level can be even higher than the signal itself. An alternative is to explore the slope changes which are observed within QRS complex edges, such that one identifies an increasing slope at the initial edge and a decreasing slope ate the final edge.

In general, we depart from an already automatically detected reference location: the R-wave peak or the most enhanced peak within the QRS complex, and the corresponding peaks for P-wave and T-wave. Then after defining searching windows centered at the corresponding fiducial points of the ECG characteristic waves, if we apply an algorithm for baseline wandering removal, and the first derivative over the resultant signal, extreme values or critical points are associated to local maxima or minima slopes within the original signal, and zero-crossings are associated to local maxima or minima within the original signal. Therefore a threshold computed within the differentiated ECG from the amplitudes associated with the first local maximum or minimum slope and the last local maximum or minimum slope, respectively, within the original signal, allows us to estimate the start point of the Q-wave and also the end point of the S-wave, concerning QRS segmentation. An analogous

Developments and Applications for ECG Signal Processing. https://doi.org/10.1016/B978-0-12-814035-2.00011-6

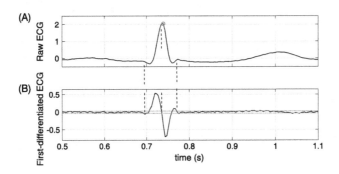

FIGURE 5.1

Signal window around the first detected QRS complex and the corresponding estimations
for QRS onset and end within the record sel103/qtdb: (A) a complete ECG cycle and its
R-wave peak identified with a circle, and (B) the corresponding first-differentiated resultant
signal with thresholds computed from the amplitudes associated with the first local
minimum slope and the last local maximum slope within the original signal, respectively.

analysis may be applied for estimating the locations of start and end points for P- and
T-waves.

As well emphasized by Sörnmo and Laguna (2005), the above search procedure
for delineating QRS complex considers that each of the regular component waves
is present (Q-wave, R-wave and S-wave). As according to the QRS morphology under study, a lower (sometimes only one) or larger number of component waves are
present, it is necessary firstly to detect which are the present/absent waves to ensure
meaningful and accurate delineation. Such identification may be implemented, for
example, by counting the number of critical points and verifying their ordination and
signs in an interval positioned around the QRS fiducial points, and associating the
patterns of successive peak sign/amplitudes and interpeak intervals with morphology
classes (for analysis of the characteristics related to different QRS morphologies, see
Lugovaya, 2005).

As illustrated by Fig. 5.1, which exemplifies the estimation of QRS onset and end
for record sel103/qtdb (pertaining to QT Database Laguna et al., 1997), the threshold
level, which helps to estimate the position of a wave edge, may be computed as a
certain percentage of the maximum slope or may be computed with reference to a
slope value representative of the boundary to be determined. As the predominant
QRS morphology of the record sel103/qtdb presents all the regular individual waves,
we may expect to find four critical points, the first one associated with the beginning
of the Q-wave and the last one associated with the end of the S-wave.

Considering the presence of broad-band spectral content noise within real ECG
signals, wave delineation from the differentiated ECG provides poorly accurate results, since the noise at QRS onsets and ends may present amplitudes higher than
the signal itself. And also, the first derivative filter works as a high-pass filter,
which enhances noise with high-spectral content. Concerning the attenuation of high-
frequency noise, it is expected that the performance of a QRS delineator algorithm

based on the first derivative achieves better results by inserting a low-pass filter, whose cutoff frequency may be adapted to the spectral content of the wave to be delineated, example, the predominant QRS morphology. The difficulty of applying QRS delineating based on classic filtering is that the spectral content of the QRS complex changes from patient to patient and even for a single patient. It is well known that some cardiac conditions are related to widen QRS complexes or notched waves, whose spectral content differs significantly from that related to a regular morphology.

According to Sörnmo and Laguna (2005), an alternative for obtaining QRS delineation algorithms more robust to noise is to apply delineating approaches based on template matching. The family of approaches based on template matching seeks a mathematical model, which is highly correlated with a given QRS morphology or with the lowpass differentiated ECG for posterior estimation of QRS boundaries. Thus these approaches do not use just local information (threshold levels based on maximum slopes), but the whole QRS waveform. Some methodologies that apply template matching for QRS delineation are: Madeiro et al. (2007), Krasteva and Jekova (2007), Singh and Gupta (2009).

Despite we do not have universal or golden rules for evaluating QRS delineation, the parameters of the applied/chosen approaches should be tuned, such that the estimation for QRS onsets and ends agrees (in a certain degree) with those obtained by medical expertise, by comparing within a reference and annotated database. A well-known database, QT Database, publicly accessed through PhysioNet website, provides expertise annotations for QRS onset/ends (Goldberger et al., 2000; Laguna et al., 1997). As widely applied and found in the literature, delineation performance measures are computed by measuring the mean and the standard deviation of the time differences (or errors) between the respective estimations for QRS edges (time locations) and the locations suggested by the expertise. Considering that two or more experts rarely agree about the exact location of waveform boundaries, a zero value of the mean or standard deviation of the time differences is not expected. According to Sörnmo and Laguna (2005), a set of results for a given approach is judged as satisfactory when the standard deviation of the time differences, computed for a given waveform boundary estimation, considering a given manual annotation reference, is approximately in the same order as that standard deviation derived from the differences among experts' suggestions (Party, 1985).

In the next section, we will describe and evaluate a group of different approaches for QRS delineation under different scenarios, including noise interference, physiological noise, different arrhythmia classes and QRS morphology alterations, different leads from the same record, and also evaluate computational approaches for solving a broad range of problems/challenges. We will also rank a subset of the most-cited techniques concerning the mean and the standard deviation of the time differences between automatic estimations and expert annotations, considering QT Database (PhysioBank) annotation marks.

5.2 ANALYZING DIFFERENT APPROACHES FOR QRS DELINEATION

The approach developed by Madeiro et al. (2012) is characterized by some adaptations to the one developed by Illanes-Manriquez and Zhang (2008). Both approaches obtain a QRS envelope based on band-pass filtering and Hilbert transform. The technique proposed by Madeiro et al. (2012) implements band-pass filtering using Continuous Wavelet Transform and also considers first derivative filter. After obtaining a signal envelope within a signal interval containing each QRS complex, the algorithms compute a surface indicator $A(t)$ for a given QRS waveform. The surface indicator refers to the area covered by each QRS envelope inside a moving window.

Each signal interval containing a QRS complex is not exactly centered at each R-wave. It is defined by two time offsets, considering $[R_p - 300 \text{ (ms)}, R_p]$ for searching QRS onset and $[R_p, R_p + 190 \text{ (ms)}]$ for searching QRS offset, where R_p is the estimation for the location of the R-wave peak.

Naming the resultant envelope signal as $V(n)$, and considering that the start point (t_1) and the end point (t_2) of a QRS envelope are associated with QRS onset and offset, respectively, and that $L = t_2 - t_1$ is the QRS envelope length, both approaches aim at maximizing the following expressions

$$A_1(t) = \int_{t-WL}^{t} [V(\tau) - V(t)]d\tau, \tag{5.1}$$

$$A_2(t) = \int_{t}^{t+WL} [V(\tau) - V(t)]d\tau. \tag{5.2}$$

It can be shown that $A_1(t)$ and $A_2(t)$ are, respectively, maximized at the instants t_2 (QRS offset) and t_1 (QRS onset).

Each integral is computed as a summation in a moving window and understood as the area under the envelope $V(\tau)$, above a horizontal reference crossing $(t, V(t))$ in the interval $[t - WL, t]$, for searching QRS offset, and $[t, t + WL]$ for searching QRS onset.

According to Illanes-Manriquez and Zhang (2008), due to QRS morphology changes, the value of the parameter WL should be adapted for each analyzed QRS waveform. Considering t_p as the location of the maximum of the envelope $V(n)$, the parameter WL should satisfy $t_2 - t_p < WL < L$, for QRS offset estimation, and it should satisfy $t_p - t_1 < WL < L$ for QRS onset estimation. As we do not know the instants t_1 and t_2 a priori, initially, the metrics $A_1(t)$ and $A_2(t)$ are computed using $WL = WL_0$, which is defined as the maximum duration for a QRS complex encountered in practice (set as 150 ms). Then we detect a first estimation for QRS offset, $t = s_2$, and a first estimation for QRS onset, $t = s_1$. We also find the location of the envelope peak t_p, which does not depend on WL value. Then we update WL value as $WL = s_2 - t_p$ for QRS offset estimation, and as $WL = t_p - s_1$ for QRS onset estimation.

Considering that the increase of the scale factor within Wavelet transform results in a decreased time resolution within the resultant filtered signals, and that some QRS morphologies (possibly related to cardiac diseases) are widened and present additional low frequency spectral content not found at regular QRS morphologies, beyond adapting the parameter WL, the approach developed by Madeiro et al. (2012) also proposes some conditional tests for adapting wavelet transform scale factor, among the possible values: $2^0, 2^1, 2^2$, and 2^3. Depending on the result of the conditional tests, it may be possible that the algorithm derive a resultant envelope version, specific for QRS onset estimation, and another version specific for QRS offset estimation.

On the contrary, the approach proposed by Illanes-Manriquez and Zhang (2008) applies bandpass filtering with fixed parameters for automatic estimation of QRS onset (cutoff frequencies 0.5–40 Hz) and QRS offset (cutoff frequencies 5–30 Hz). The implemented filter is based on the computation of the Fast Fourier Transform of the input ECG signal, truncation of the frequencies outside the targeted bands and the computation of the inverse Fourier Transform.

Figs. 5.2(A)–(D) illustrate the process of obtaining QRS envelope, by applying successively Wavelet transform (scale factor $= 2^1$), first derivative, and Hilbert transform over windows around already detected QRS complexes within the records sel103 and sel114 (channel 1 and channel 2), respectively. The samples related to the beginning and end of the corresponding QRS envelopes are detected through the surface indicator $A(t)$, which is computed over the envelopes (Eqs. (5.1) and (5.2)). We can observe small differences/deviations between the automatic estimations applied over different leads/channels for a single record, deviations more evident for QRS offset.

As for illustrating performance comparison, Figs. 5.3(A)–(D) present the process of obtaining QRS envelope, by applying band-pass filtering, using only Fast Fourier Transform, truncation of the frequencies outside the targeted bands and the computation of the inverse Fourier Transform, and Hilbert transform, over windows around already detected QRS complexes (approach proposed by Illanes-Manriquez and Zhang, 2008), respectively, within the records sel103 and sel114 (channel 1 and channel 2). The samples related to the beginning and end of the corresponding QRS envelopes are detected through the surface indicator $A(t)$, as applied by Madeiro et al. (2012), which is computed over the envelopes (Eqs. (5.1) and (5.2)). We can observe the Wavelet transform acting as band-pass filtering, and that the insertion of the first derivative before computing the signal envelope (approach developed by Madeiro et al., 2012), provide an increasing of the time accuracy (see Figs. 5.2).

The approach developed by Martínez et al. (2004) also departs from an interval containing the already detected R-wave peak and applies the criteria of slope threshold within wavelet transform of the input signal. As we detailed in the previous chapter, the algorithm associates the zero-crossing of the wavelet transform at scale 2^1, between a positive maximum-negative minimum pair, to a QRS fiducial point (QRS occurrence). It is expected that the same zero-crossing associated with R-wave peak be located between critical points with opposite signs at scale 2^2. If we denote the first of these critical points (located immediately before the zero-crossing) as n_{pre} and the second one as n_{pos} (located immediately after the zero-crossing), the algorithm searches before n_{pre} and after n_{pos} for significant maxima of $|W_{2^2}x[n]|$, that

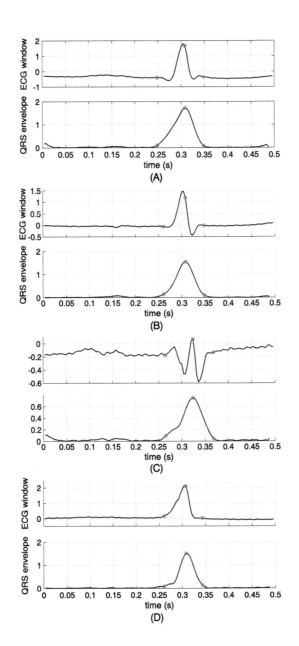

FIGURE 5.2

Delineation of QRS complex through the computation of the QRS envelope for different QRS morphologies considering the approach proposed by Madeiro et al. (2012): (A) qRs morphology within record sel103/qtdb (first channel), (B) Rs morphology within record sel103/qtdb (second channel), (C) rsR's' morphology within record sel114/qtdb (first channel) and (D) R morphology within record sel114/qtdb (second channel).

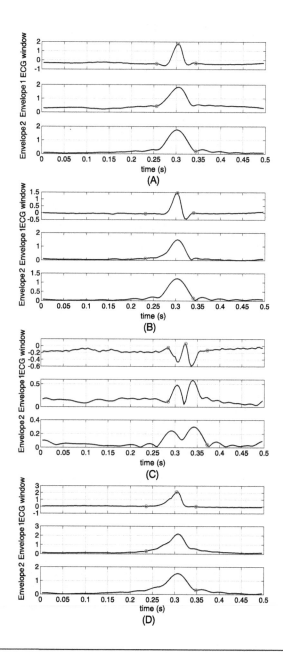

FIGURE 5.3

Delineation of QRS complex through the computation of the QRS envelope for different QRS morphologies considering the approach proposed by Illanes-Manriquez and Zhang (2008): (A) qRs morphology within record sel103/qtdb (first channel), (B) Rs morphology within record sel103/qtdb (second channel), (C) rsR's' morphology within record sel114/qtdb (first channel) and (D) R morphology within record sel114/qtdb (second channel).

is, maxima or minima critical points within the scaled wavelet transform $W_{2^2}x[n]$. Two thresholds are computed for identifying maxima or minima critical points as significant: $\gamma_{QRS_{pre}}$ or $\gamma_{QRS_{pos}}$. For a given search window, the thresholds are computed as

$$\gamma_{QRS_{pre}} = 0.06 \max(|W_{2^2}x[n]|), \tag{5.3}$$

$$\gamma_{QRS_{pos}} = 0.09 \max(|W_{2^2}x[n]|), \tag{5.4}$$

where n refers to the time indices pertaining to each search window.

Once identified, all the significant critical points associated with a given QRS complex, some of them located before the R-wave and others after the R-wave, we detect the first and the last critical points associated with a given QRS, named n_{first} and n_{last}. Thus the QRS onset should be identified before n_{first}, and the QRS offset should be identified after n_{last}. According to Martínez et al. (2004), two candidates are identified for QRS onset. The first one is the sample, where the modulus $|W_{2^2}x[n]|$ is below a threshold, named $\xi_{QRS_{on}}$, computed as

$$\xi_{QRS_{on}} = \begin{cases} 0.05\,W_{2^2}x[n_{first}] \text{ if } W_{2^2}x[n_{first}] > 0, \\ 0.07\,W_{2^2}x[n_{first}] \text{ if } W_{2^2}x[n_{first}] < 0. \end{cases} \tag{5.5}$$

The second candidate is the first local minimum of $|W_{2^2}x[n]|$, detected before n_{first}. We take as QRS onset the candidate located at the nearest point to the QRS fiducial point (zero-crossing between n_{pre} and n_{pos}).

Analogously, two candidates for QRS offset are identified. The first one is the sample, where the modulus $|W_{2^2}x[n]|$ is below a threshold, named $\xi_{QRS_{off}}$, computed as

$$\xi_{QRS_{off}} = \begin{cases} 0.125\,W_{2^2}x[n_{last}] \text{ if } W_{2^2}x[n_{last}] > 0, \\ 0.71\,W_{2^2}x[n_{last}] \text{ if } W_{2^2}x[n_{last}] < 0. \end{cases} \tag{5.6}$$

The second candidate is the first local minimum of $|W_{2^2}x[n]|$, detected after n_{last}. We take as QRS offset the candidate located at the nearest point to the QRS fiducial point (zero-crossing between n_{pre} and n_{pos}).

It is evident that the threshold parameters suggested by Martínez et al. (2004) are tuned after exhaustive computing simulations, and it may be necessary to adapt these parameters according to the sampling frequency, noise interference, and QRS morphology changes due to specific cardiac diseases or medical procedures/interventions. Also some protection measures should be taken, as in regards to the maximum temporal distance between n_{first} and the QRS fiducial point, and between n_{first} and the QRS fiducial point.

For performance comparison, Figs. 5.4(A)–(F) present the process of obtaining wavelet transform $W_{2^2}x[n]$, and associating the QRS onset and offset within the ECG window, respectively, to samples located before the first significant critical point and after the last significant critical point within the scaled wavelet transform of the input signal window (approach proposed by Martínez et al., 2004), using the records sel103, sel114, and sel221 (channel 1 and channel 2). We can observe

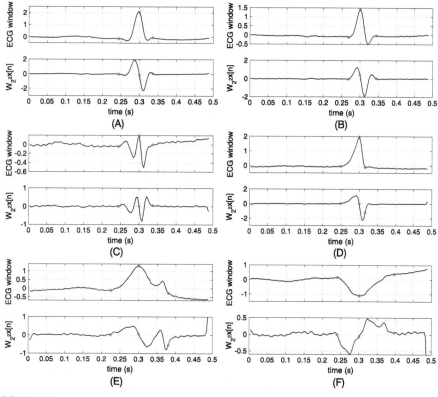

FIGURE 5.4

Delineation of QRS complex through the computation of the wavelet transform $W_{2^2}x[n]$ for different QRS morphologies, considering the approach proposed by Martínez et al. (2004): (A) qRs morphology within record sel103/qtdb (first channel), (B) Rs morphology within record sel103/qtdb (second channel), (C) rsR's' morphology within record sel114/qtdb (first channel), (D) R morphology within record sel114/qtdb (second channel), (E) Rr' morphology within record sel221/qtdb (first channel), and (F) QS morphology within record sel221/qtdb (second channel).

the robustness of the referred approach, which provides accurate results for different QRS morphologies, including examples of premature ventricular contraction (Figs. 5.4(E)–(F)), which present widened and notched QRS complexes. For detecting QRS onset and offset, the approach applies flexibility since we can find the points using threshold criteria or searching for local minimum of $|W_{2^2}x[n]|$.

As we have detailed in the previous chapter, the approach proposed by Ghaffari et al. (2009) applies beyond the linear wavelet transform over the raw ECG signal, using the scale 2^3, which provides a satisfactory enhancing for QRS complex and attenuating for low-frequency and high-frequency noise, including physiologic noise, also a nonlinear transformation, represented by the time-series ACL. The metric ACL is defined by the product of a function related to the area below the waveform and a function representing a curve of the waveform. For convenience, we rewrite here the equations required for computing ACL metric,

$$y_k = W_{2^\lambda}[k : k + L], \tag{5.7}$$

where y_k is a vector, including the samples k to $k + L$ of the filtered version related to scale 2^λ. Then for each value of k, one defines the area under the absolute value of the time-series y_k and the curve length of y_k as

$$Area[k] = \int_{t_{0k}}^{t_{fk}} |y_k(t)| dt, \tag{5.8}$$

$$Curve[k] = \int_{t_{0k}}^{t_{fk}} \sqrt{1 + \dot{y}_k^2} = \sum_{t_{0k}+1}^{t_{fk}} \sqrt{1 + (y_k(n) - y_k(n - 1))^2}, \tag{5.9}$$

where t_{0k} and t_{fk} are, respectively, the start and end points of the sequence y_k.

A variable thresholding is computed throughout the ACL-time-series based on mean and standard deviation of the amplitude values. This variable threshold provides the detection of maximum values, which are associated to R-wave peaks. Here the R-waves are again used as initial references for searching QRS boundaries. After detecting maximum critical points associated with QRS fiducial points, we proceed with a search, departing from the already detected peaks towards the right side and left side until finding local minima for the absolute ACL slope. Each local minimum situated at each side from the R-wave peak reference is associated with a specific QRS boundary. Once QRS edges are detected through searching within ACL metric, it is possible to map the corresponding automatic markings within the wavelet transform used as input for ACL computing, and then to identify the number and signs of critical points, such that recognition of a given predominant QRS morphology is allowed.

Considering that ACL metric is computed using a window approximately 40–50 ms, and each sample of ACL metric requires computing area and curve metrics using a full signal window, the corresponding automatic boundaries found within the ACL metric, and even the QRS fiducial point, may not be mapped directly within the ECG signal. It is clear that there will be a time delay between the peaks (maximum values) within ACL series and the QRS peaks, and also between the boundaries from the enhanced waves within ACL and the real QRS edges. That is due to the forward computing of ACL. Also we have the issue of loss time resolution as a result of increasing wavelet transform scale factors, and the authors employment of scale 2^3 as input signal for computing ACL. The authors do not detail how to solve these issues. A possible solution, which should be tested trough computational simulations, is to transfer the detected points (peaks and boundaries) from the ACL signal to wavelet transform within a predetermined scale and to proceed a forward searching for zero-crossings and minimum values (associated, respectively, with peaks and edges within the original signal).

Furthermore, some protection measures should be taken, such as using maximum time deviations between QRS fiducial points and the automatic detections of boundaries. As the proposed approach suggests to associate QRS edges to local minima

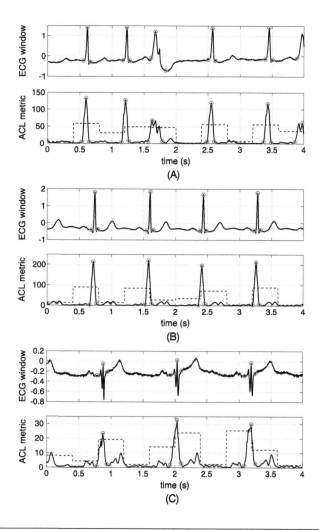

FIGURE 5.5

Delineation of QRS complex through the computation of the Area-Curve length metric considering the approach proposed by Ghaffari et al. (2009) for different morphologies. (A) Delineating QRS complex through ACL metric: record sel221/qtdb (first channel). (B) Delineating QRS complex through ACL metric: record sel103/qtdb (first channel). (C) Delineating QRS complex through ACL metric: record sel114/qtdb (first channel).

for absolute ACL slope, it may occur that a real local minimum appears so far as the QRS peak. In these situations, we may apply threshold for ACL slope, or even zero-crossing.

Figs. 5.5(A)–(C) illustrate the process of obtaining ACL metric (40-ms window), by applying successively Wavelet transform (scale factor $= 2^3$) and computing the metrics in Eqs. (5.7) to (5.9) within the records sel221, sel103, and sel114 (channel 1).

The samples detected as maximum critical points within ACL metrics are transferred to the wavelet transform (scale 2^1) and a forward searching for zero-crossing is commenced. For mapping the detected QRS boundaries within the original, each edge detected within the ACL signal is shifted by 40 ms.

5.3 NEW TRENDS AND DEVELOPMENTS

One promising innovative technique, recently proposed and still being tested and developed consists of searching a template matching between different QRS complex morphologies and combined and distorted mathematical functions (Caldas et al., 2017).

It is well known that computing mathematical modeling for ECG waveforms is by no means new. Richardson et al. (1971) proposed three mathematical models based on Gaussian functions and their first and second derivatives for modeling QRS complex, P-wave, and T-wave. Initially the goal was to subsidize the spectral analysis of the real cardiac signal, departing from a fitted function. One of the great advantages of this approach is that the modeled signal is free of noise. Furthermore, more recent developments have applied mathematical modeling for waveform characterization, that is, beyond the traditional temporal metrics, which are extracted by the classic waveform delineators, amplitude, and durations; those approaches also provide features regarding waveform morphology, such as asymmetry and increasing or decreasing of distortions (Wohlfart, 1987; Malik and Camm, 1989; Padrini et al., 1995; Madeiro et al., 2013).

Having in mind that real ECG waveforms are intrinsically asymmetric and even distorted by interference and physiological noise, it is intuitive that mathematical functions should acquire a certain degree of distortion for being able to fit physiological signal. This idea was applied by Madeiro et al. (2017) for modeling T-waves and P-waves, through the combining of Gaussian and Rayleigh functions. However, the QRS complex presents a considerable number of typically occurring morphologies, which make its mathematical modeling challenging. As already detailed, Lugovaya (2005) list twenty common QRS morphologies concerning a study for biometric human identification based on ECG: qR, qRs, Rs, R, RS, rSR', rR', qrSr', RSr', rR's, rS, rSr', Qr, QS, QR, qrS, qS, rsR's', QRs, and Qrs. Taking as reference this set of morphologies, Caldas et al. (2017) proposed applying five mathematical models based on adjustments, and combining three known mathematical functions: the Gaussian function, the Mexican-hat function (Gaussian second-derivative), and the Rayleigh function (derived from Rayleigh probability density function).

Thus the first mathematical model corresponds to the combination of two Gaussian functions. A single discrete-time Gaussian function may be defined as (Madeiro et al., 2017)

$$G_{\sigma_1}[k] = \frac{1}{\sqrt{2\pi}\,\sigma_1}\exp(\frac{k^2}{2\sigma_1^2}), \qquad (5.10)$$

where k varies within a given interval $-x_1 \leq k \leq x_2$, and the parameter σ_1 is related to the Gaussian width.

The mathematical model resultant from combining two discrete time Gaussian functions with individual width parameters σ_1 and σ_2 is therefore given as

$$G_{\sigma_1,\sigma_2}[k] = \begin{cases} G_{\sigma_1}[k] \text{ if } -x_1 \leq k \leq 0, \\ G_{\sigma_2}[k] \text{ if } 0 \leq k \leq x_2. \end{cases} \tag{5.11}$$

The second mathematical model corresponds the second derivative of the first mathematical model, that is, a modified Mexican-hat function, once the original one is symmetric. Therefore given $G_{\sigma_1,\sigma_2}[k]$, derived from the combination of two Gaussian functions, the second mathematical model $G_{\sigma_1,\sigma_2}[k]^d$ may be obtained as a second-order progressive divided difference

$$G^d_{\sigma_1,\sigma_2}[k] = \frac{G_{\sigma_1,\sigma_2}[k+1] - 2G_{\sigma_1,\sigma_2}[k] + G_{\sigma_1,\sigma_2}[k-1]}{T_s^2}, \tag{5.12}$$

where T_s represents the sampling period.

Finally we define a discrete time Rayleigh probability density function $R_{\sigma_1}[k]$ within the interval $0 \leq k \leq h$:

$$R_{\sigma_1}[k] = \frac{k}{\sigma_1^2} \exp(-\frac{k^2}{2\sigma_1^2}), \tag{5.13}$$

where σ_1 is the corresponding variable parameter.

Then the next three mathematical models correspond to three different combinations of two Rayleigh functions: Rayleigh cycle positive/negative, Rayleigh cycle negative/positive, and Rayleigh cycle positive/positive.

Concerning the third mathematical model, we obtain a composition of two Rayleigh functions, denoted as $R_{\sigma_1,\sigma_2}[k]$, $0 \leq k \leq 2h$. We depart from the definition of two discrete time Rayleigh probability density functions $R_{\sigma_1}[k]$ and $R_{\sigma_2}[k]$ so that

$$R_{\sigma_1,\sigma_2}[k] = \begin{cases} R_{\sigma_1}[h-k] \text{ if } 0 \leq k \leq h, \\ -R_{\sigma_2}[k-h] \text{ if } h \leq k \leq 2h. \end{cases} \tag{5.14}$$

The other two mathematical models correspond to variations of the third mathematical model, the fourth one (Rayleigh cycle negative/positive) given as

$$R_{\sigma_1,\sigma_2}[k] = \begin{cases} -R_{\sigma_1}[h-k] \text{ if } 0 \leq k \leq h, \\ R_{\sigma_2}[k-h] \text{ if } h \leq k \leq 2h \end{cases} \tag{5.15}$$

and the fifth one (Rayleigh cycle positive/positive) given as

$$R_{\sigma_1,\sigma_2}[k] = \begin{cases} R_{\sigma_1}[h-k] \text{ if } 0 \leq k \leq h, \\ R_{\sigma_2}[k-h] \text{ if } h \leq k \leq 2h. \end{cases} \tag{5.16}$$

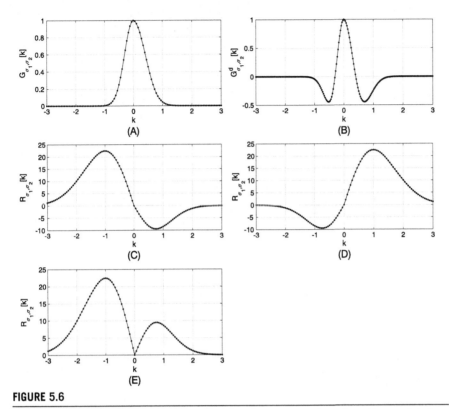

FIGURE 5.6

Illustrative examples of the five mathematical functions proposed as alternative models for fitting QRS morphologies. (A) First model: combining two discrete-time Gaussian functions. (B) Second model: modified Mexican-hat obtained by the second derivative of the first model. (C) Third model: combining two Rayleigh probability density functions (Rayleigh cycle positive/negative). (D) Fourth model: combining two Rayleigh probability density functions (Rayleigh cycle negative/positive). (E) Fifth model: combining two Rayleigh probability density functions (Rayleigh cycle positive/positive).

Figs. 5.6(A)–(E) illustrate samples of each one of the proposed mathematical models for QRS complex fitting.

After detecting each QRS complex within a given ECG record, it is possible to establish search windows centered in each QRS fiducial point (R-wave peak) with the aim to investigate which model is more appropriate for a given predominant morphology. Through an exhaustive effort, we managed to fit each model within a given morphology and computed a normalized RMS error between a given QRS segment and a given fitted model. The successful mathematical model is the one which allows obtaining the lowest normalized RMS error, which may be computed as

$$\varepsilon = \sqrt{\frac{\sum_{k=1}^{L} |W_X[k] - \hat{W}_X[k]|^2}{\sum_{k=1}^{L} W_X[k]^2}}, \tag{5.17}$$

where $W_X[k]$ refers to the QRS segment, for which we desire to fit a model, and $\hat{W}_X[k]$ refers to the evaluated kernel.

Figs. 5.7(A)–(H) illustrate examples of QRS modeling within the records 100, 103, 105, and 118 from MIT–BIH Arrhythmia Database, and a boxplot analysis of the normalized RMS errors related to each mathematical model for fitting each QRS morphology. In this sense, Fig. 5.7(A) presents the second proposed mathematical function (modified Mexican-hat) as the best fitted function for modeling the qRs morphology within record 100/mitdb. According to the boxplot analysis presented by Fig. 5.7(B), we can infer that qRs morphology prevails as predominant QRS morphology within record 100/mitdb. Fig. 5.7(C) shows the third proposed mathematical function (composition of two Rayleigh functions, cycle positive/negative) as the best fitted function for modeling the Rs morphology within record 103/mitdb. Analogously, the boxplot analysis presented by Fig. 5.7(D) indicates that RS morphology prevails as predominant QRS morphology within record 103/mitdb. As for Figs. 5.7(E) and 5.7(F), despite the composition of two Gaussian functions fit satisfactorily the QRS morphology within record 105/mitdb, the boxplot analysis of the normalized RMS errors shows that the successful mathematical model, that is, the mathematical model, which allows achieving the lowest values for RMS error, alternates among the five candidates, which may indicate intense QRS morphology variations. Finally Fig. 5.7(G) presents the fourth proposed mathematical function (composition of two Rayleigh functions, cycle negative/positive) as the best fitted function for modeling the qR morphology within record 118/mitdb. According to the boxplot analysis presented by Fig. 5.7(H), we can infer that qR morphology prevails as predominant QRS morphology within record 100/mitdb.

5.4 EVALUATING AND COMPARING PERFORMANCES OF QRS DELINEATORS

As already explained, the approaches for QRS delineating are evaluated by computing the time differences between automatic detections for QRS edges and manual expertise annotations. It is important to emphasize that most of the approaches available in literature work on a single channel basis, that is, they do not consider multiple detections using several channels for correcting eventual discrepancy and establishing a unique reference. Considering that experts perform manual annotation process, having in sight several leads (all available ones), published approaches apply the delineation algorithms over each individual channel and obtain two or more sets of results, that is, one per channel. Then for each wave edge, the time error is computed considering the channel (the specific detection), which allows obtaining the lowest error. However, when a relatively high number of leads are considered within an ECG record, a more complex rule may be required. For example, Martínez et al. (2004) applied their Wavelet-based delineator over the Common Standards for Electrocardiography (CSE) multilead measurement database, for which each record contains 15 available leads. Therefore after obtaining 15 different sets of automatic detec-

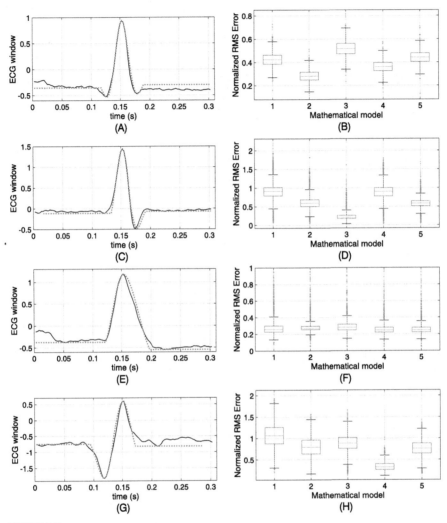

FIGURE 5.7

Applying mathematical models for QRS modeling: analysis of fitting and boxplot of results of normalized RMS errors concerned each individual mathematical model for fitting each QRS morphology. (A) Record 100/mitdb: QRS modeled by modified Mexican-hat function. (B) Record 100/mitdb: boxplot analysis of normalized RMS errors for fitting QRS morphology. (C) Record 103/mitdb: QRS modeled by combining two Rayleigh probability density functions (cycle positive/negative). (D) Record 103/mitdb: boxplot analysis of normalized RMS errors for fitting QRS morphology. (E) Record 105/mitdb: QRS modeled by combining two Gaussian functions. (F) Record 105/mitdb: boxplot analysis of normalized RMS errors for fitting QRS morphology. (G) Record 118/mitdb: QRS modeled by combining two Rayleigh probability density functions (cycle negative/positive). (H) Record 118/mitdb: boxplot analysis of normalized RMS errors for fitting QRS morphology.

tions, the authors ordered them (each edge detection) and selected as the beginning (end) of a wave the first (last) detected location, whose k nearest neighbors lay within a δ-ms interval.

Also for some databases, such as QT Database (QTDB), more than one manual annotation reference set is available, which allows computing the delineation deviations for a given technique with respect to each expert set and also the differences (mean and standard deviation) between referees (interexpert differences). A criterion for considering an automatic approach as satisfactory is to check if the standard deviation of the time errors between the detected wave edge locations and each referee annotation set remains at values with the same order as that standard deviation derived from interexpert differences (Martínez et al., 2004; Madeiro et al., 2012). Therefore the standard deviations of the errors (repeatability) are more important and decisive for evaluating a given approach than the simple mean of the errors.

From among the databases containing standard manual annotations, mostly the QT Database is applied (Kohler et al., 2002; Laguna et al., 1997). The QT Database was built for providing means to evaluate algorithms that perform waveform delineation within the ECG, that is, for QRS complex, P-wave, and T-wave segmentation. Altogether 105 records are available with a 250-Hz sampling frequency, and the set of records provides a broad variety of waveform morphologies, and also pathologic conditions, and noise or interference incidence. Each ECG signal presents a length of 15 minutes and two variable channels for signal processing, and has been selected from other databases, such as MIT–BIH Arrhythmia Database and the European ST–T Database. QT Database provides expert annotations for at least 30 beats per recording, with labels including peaks of the QRS complexes, P-, and T-waves, and also the onsets and ends. Concerning the evaluation for QRS delineation, the database provides a total of free 2909 annotations (Madeiro et al., 2012).

Another frequently applied database for evaluating QRS delineation algorithms is the Common Standards for Electrocardiography (CSE) multilead measurement database (Laguna et al., 1997), the CSEDB. This dataset provides around 1000 multichannel recordings (12 or 15 leads) with a sampling frequency of 500 Hz (Kohler et al., 2002). According to Martínez et al. (2004), the number of manually annotated beats is scarce, around 32 beats. Concerning the available annotated beats, CSEDB provides median referee annotations, which were computed using data supplied by five expert cardiologists.

In order to make a comparison study, we rank a subset of the most current reported results regarding the algorithms tested against QT Database, as synthesized within Table 5.1, and the algorithms tests against CSE Database, as synthesized within Table 5.2.

From the listed results, considering the 250-Hz sampling frequency of QTDB, the mostly applied database, all the algorithms have a standard deviation of the errors of approximately 2 or 3 samples. And all the algorithms obtained average errors around or lower than one sampling interval. For the 500-Hz sampling frequency of CSEDB, each single approach achieves more accurate/stable results, which may be justified

Table 5.1 Performance comparison of several QRS delineation algorithms: application to QTDB. (NR: Not reported)

QRS delineation algorithm	Annotations	Onset deviation (ms)	Offset deviation
Madeiro et al. (2012)	2909	2.85 ± 9.90	2.83 ± 12.26
Martínez et al. (2010)	3623	−0.2 ± 7.2	2.5 ± 8.9
Martínez et al. (2004)	3623	4.6 ± 7.7	0.8 ± 8.7
Ghaffari et al. (2009)	3623	−0.6 ± 8.0	0.3 ± 8.8
Madeiro et al. (2007)	2909	−3.4 ± 11.6	−6.5 ± 11.3
Laguna et al. (1994)	3623	−3.6 ± 8.6	−1.1 ± 8.3
Di Marco and Chiari (2011)	3623	−5.1 ± 7.2	0.9 ± 8.7

Table 5.2 Performance comparison of several QRS delineation algorithms: application to CSEDB. (NR: Not reported)

QRS delineation algorithm	Annotations	Onset deviation (ms)	Offset deviation
Ghaffari et al. (2009)	32	−0.9 ± 2.9	4.2 ± 4.7
Martínez et al. (2004)	32	1.3 ± 6.3	5.8 ± 10.9
Laguna et al. (1994)	30	−2.1 ± 7.4	−0.2 ± 3.6
De Chazal and Celler (1996)	32	0.9 ± 3.6	−0.6 ± 7.1
Vítek et al. (2010)	NR	0.3 ± 4.0	0.8 ± 4.7
Sahambi et al. (1997)	NR	NR ± 2.0	NR ± 4.0

by the higher temporal resolution, and the fact that CSEDB provides more simple signals in comparison to QTDB (Ghaffari et al., 2009).

According to Jane et al. (1997), the mean error is analyzed to evaluate how close is the approach's criterion to the expert's criterion. On the other hand, the standard deviation is concerned within the stability obtained by the analyzed approach. Some accepted standard deviation tolerances are widely considered and follow the recommendations of the CSE working party (Party, 1985), which are derived from the measurements made by different experts. In this sense, the acceptable standard deviation error values for QRS onset and offset are, respectively, 6.5 ms and 11.6 ms. Sörnmo and Laguna (2005) suggest an acceptable average error as 15 ms. However, it is important to emphasize that the interpretation of the CSE tolerances when assessing the results of QRS delineation is not so simple, since they were computed from a set of records with different number of channels, resolutions, sampling frequency, quality, and rhythms (Martínez et al., 2004).

Additionally, Beraza and Romero (2017) and Jane et al. (1997) proposed to classify the output of a given algorithm when applied within a given ECG record into one of 4 groups, depending on the obtained parameters μ and σ, mean and standard deviation of the errors: Group 1, acceptable μ and σ; Group 2, acceptable σ but nonacceptable μ; Group 3, acceptable μ but nonacceptable σ; and Group 4, nonacceptable μ and σ. Then for each group, one associates a characteristic concerned with the processed exam. Thus within group 1, we identified ECG signals with reasonable detections; group 2 is commonly associated with morphology identification errors (systematic errors), group 3 is commonly associated with low SNR or small-amplitude waves, and group 4 combines situations of groups 2 and 3, that is, morphology identification error with low SNR.

Despite the small differences regarding the performance obtained by several QRS delineators, we may say that the current research status for automatic QRS segmentation issue already achieves almost negligible errors. However, we can highlight some still highly active research demands having to do with QRS feature extraction. Beyond the classical parameters (amplitude, polarity, and duration of QRS complex), which characteristics related to the morphology may be useful for recognizing or predicting important events? How to recognize subtle changes within waveform, which do not necessarily impact amplitude and duration? Is it possible to condense multi-channel information in a single information signal? Furthermore, as for QRS detection, the most current QRS delineators require sufficient computational resources. Are these methods fast and feasible in real-time applications? Are these methods applicable for smartphone environment and other battery-driven devices? In this context, important demands for research are the mechanisms for acquiring and recognizing subtle changes within QRS complex and to increase the efficiency of the algorithms, adapting them for real-time applications and battery-driven devices.

5.5 CONCLUSIONS

In this chapter, we presented basic concepts related to edge detection, which may be applied for QRS delineation. We demonstrated that QRS waveform may present several specific morphologies and that preprocessing techniques are required for emphasizing slope variation at the boundaries related to QRS onset and offset. Wavelet transforms and Hilbert transforms, when applied together, beyond providing band-pass filtering for emphasizing QRS subtle changes, also allows computing QRS envelope, which is a bell-shaped positive wave, whose boundaries coincide with QRS onset and offset. We also looked at new trends and developments concerning QRS extraction features, based on mathematical modeling. Thus it is possible, beyond monitoring evolution of classical parameters, such as QRS amplitude and duration, to observe subtle changes within QRS morphology. Finally, by analyzing the performance obtained by several QRS delineators, we may say that the current research status for automatic QRS segmentation issue already achieves almost negligible errors. However, we highlighted research yet in high, such as the search for methods fast and feasible in real-time applications and mechanisms for recognizing subtle changes within QRS complex morphology.

REFERENCES

Beraza, I., Romero, I., 2017. Comparative study of algorithms for ECG segmentation. Biomedical Signal Processing and Control 34, 166–173.

Caldas, W.L., Madeiro, J.P.V., Gomes, J.P.P., 2017. Extração de atributos para classificação de morfologias em sinais ECG. In: André, S. (Ed.), Oitavo Simpósio de Instrumentação e Imagens Médicas e Sétimo Simpósio de Processamento de Sinais, vol. 1.

De Chazal, P., Celler, B., 1996. Automatic measurement of the QRS onset and offset in individual ECG leads. In: Engineering in Medicine and Biology Society, 1996. Bridging Disciplines for Biomedicine. Proceedings of the 18th Annual International Conference of the IEEE, vol. 4. IEEE, pp. 1399–1400.

Di Marco, L.Y., Chiari, L., 2011. A wavelet-based ECG delineation algorithm for 32-bit integer online processing. Biomedical Engineering OnLine 10 (1), 23.

Ghaffari, A., Homaeinezhad, M., Akraminia, M., Atarod, M., Daevaeiha, M., 2009. A robust wavelet-based multi-lead electrocardiogram delineation algorithm. Medical Engineering & Physics 31 (10), 1219–1227.

Goldberger, A.L., Amaral, L.A., Glass, L., Hausdorff, J.M., Ivanov, P.C., Mark, R.G., Mietus, J.E., Moody, G.B., Peng, C.K., Stanley, H.E., 2000. PhysioBank, PhysioToolkit, and PhysioNet. Circulation 101 (23), e215–e220.

Illanes-Manriquez, A., Zhang, Q., 2008. An algorithm for robust detection of QRS onset and offset in ECG signals. In: Computers in Cardiology 2008. IEEE, pp. 857–860.

Jane, R., Blasi, A., García, J., Laguna, P., 1997. Evaluation of an automatic threshold based detector of waveform limits in holter ECG with the QT Database. In: Computers in Cardiology 1997. IEEE, pp. 295–298.

Kohler, B.U., Hennig, C., Orglmeister, R., 2002. The principles of software QRS detection. IEEE Engineering in Medicine and Biology Magazine 21 (1), 42–57.

Krasteva, V., Jekova, I., 2007. QRS template matching for recognition of ventricular ectopic beats. Annals of Biomedical Engineering 35 (12), 2065–2076.

Laguna, P., Jané, R., Caminal, P., 1994. Automatic detection of wave boundaries in multilead ECG signals: validation with the CSE database. Computers and Biomedical Research 27 (1), 45–60.

Laguna, P., Mark, R.G., Goldberg, A., Moody, G.B., 1997. A database for evaluation of algorithms for measurement of QT and other waveform intervals in the ECG. In: Computers in Cardiology 1997. IEEE, pp. 673–676.

Lugovaya, T.S., 2005. Biometric Human Identification Based on ECG. PhysioNet.

Madeiro, J.P., Cortez, P.C., Oliveira, F.I., Siqueira, R.S., 2007. A new approach to QRS segmentation based on wavelet bases and adaptive threshold technique. Medical Engineering & Physics 29 (1), 26–37.

Madeiro, J.P., Cortez, P.C., Marques, J.A., Seisdedos, C.R., Sobrinho, C.R., 2012. An innovative approach of QRS segmentation based on first derivative, Hilbert and Wavelet Transforms. Medical Engineering & Physics 34 (9), 1236–1246.

Madeiro, J.P., Nicolson, W.B., Cortez, P.C., Marques, J.A., Vázquez-Seisdedos, C.R., Elangovan, N., Ng, G.A., Schlindwein, F.S., 2013. New approach for t-wave peak detection and t-wave end location in 12-lead paced ECG signals based on a mathematical model. Medical Engineering & Physics 35 (8), 1105–1115.

Madeiro, J.P.V., dos Santos, E.M.B.E., Cortez, P.C., Felix, J.H.S., Schlindwein, F.S., 2017. Evaluating Gaussian and Rayleigh-based mathematical models for T and P-waves in ECG. IEEE Latin America Transactions 15 (5), 843–853.

Malik, M., Camm, A.J., 1989. Computer model of cardiac repolarization processes and of the recovery sequence. Computers and Biomedical Research 22 (2), 160–180.

Martínez, A., Alcaraz, R., Rieta, J.J., 2010. Application of the phasor transform for automatic delineation of single-lead ECG fiducial points. Physiological Measurement 31 (11), 1467.

Martínez, J.P., Almeida, R., Olmos, S., Rocha, A.P., Laguna, P., 2004. A wavelet-based ECG delineator: evaluation on standard databases. IEEE Transactions on Biomedical Engineering 51 (4), 570–581.

Padrini, R., Butrous, G., Camm, A.J., Malik, M., 1995. Algebraic decomposition of the T–U wave morphology patterns. Pacing and Clinical Electrophysiology 18 (12), 2209–2215.

Party, C.W., 1985. Recommendations for measurement standards in quantitative electrocardiography. European Heart Journal 6 (10), 815–825.

Richardson, J., Haywood, L., Murthy, V., Harvey, G., 1971. A mathematical model for ECG wave forms and power spectra. Mathematical Biosciences 12 (3–4), 321–328.

Sahambi, J., Tandon, S., Bhatt, R., 1997. Using wavelet transforms for ECG characterization. An on-line digital signal processing system. IEEE Engineering in Medicine and Biology Magazine 16 (1), 77–83.

Singh, Y.N., Gupta, P., 2009. Biometrics method for human identification using electrocardiogram. In: International Conference on Biometrics. Springer, pp. 1270–1279.

Sörnmo, L., Laguna, P., 2005. Bioelectrical Signal Processing in Cardiac and Neurological Applications, vol. 8. Academic Press.

Vítek, M., Hrubeš, J., Kozumplík, J., 2010. A wavelet-based ECG delineation with improved P-wave offset detection accuracy. In: Analysis of Biomedical Signals and Images, Biosignal, vol. 20, pp. 1–6.

Wohlfart, B., 1987. A simple model for demonstration of ST–T-changes in ECG. European Heart Journal 8 (4), 409–416.

Mathematical Modeling of T-Wave and P-Wave: A Robust Alternative for Detecting and Delineating Those Waveforms

João Paulo do Vale Madeiro*, Paulo César Cortez†,
José Maria da Silva Monteiro Filho‡, Priscila Rocha Ferreira Rodrigues‡

*Institute for Engineering and Sustainable Development (IEDS), University for the International
Integration of the Afro-Brazilian Lusophony – UNILAB, Redenção, Ceará, Brazil
†Department of Teleinformatics Engineering, Federal University of Ceara, Fortaleza, Ceará, Brazil
‡Department of Computing Science, Federal University of Ceara, Fortaleza, Ceará, Brazil

6.1 BASIC CONCEPTS, MOTIVATION AND INHERENT DIFFICULTIES

Beyond the traditional time-domain metrics, the characterization of the morphology, as well as morphology changes for the ECG characteristic waves allows extracting a valuable data set conveying details of cardiac activity and events associated with cardiac anomalies, which may require emergency medical intervention. For T-wave, we can distinguish six possible morphology types: positive (+), negative or inverted (−), biphasic (+/− or −/+), upward and downward. Concerning P-wave, four possible morphologies are admitted: positive, negative and biphasic (+/− or −/+) (Sörnmo and Laguna, 2005; Martínez et al., 2004; Laguna et al., 1994; Hall, 2011).

The normal T-wave is usually asymmetric, with a slower onset (beginning) and a faster end phase, positive on almost all leads, habitually with the same polarity as QRS complex, and presenting an amplitude equivalent to about 10% to 30% of the QRS complex's (Pastore et al., 2016). Some of the T-wave abnormalities comprise: hyperacute T-waves, inverted, biphasic, camel-hump shaped, and flattened T-waves. Such occurrences may be related to several important events as: hyperkalemia (associated with positive, symmetrical and pointed T-waves), subepicardial ischemia

Developments and Applications for ECG Signal Processing. https://doi.org/10.1016/B978-0-12-814035-2.00012-8

(presence of symmetrical and pointed T-wave), changes in ventricular repolarization (flattened T-wave within left leads), myocardial infarction (ST-segment elevation), branch block (inverted T-waves), pulmonary embolism (negative T-waves), arrhythmogenic cardiomyopathy of the right ventricle (negative T-waves), hypopotassemia (signaled by biphasic T-wave), and ischemic stroke (inverted T-waves with broad base) (Surawicz and Knilans, 2008).

The normal P-wave has a maximum amplitude of 0.25 mV and a duration of less than 110 ms. Some morphology changes may occur following cardiac rhythms oscillations. The sinus rhythm of the heart (physiological rhythm) is observed in the surface ECG through the presence of positive P-waves in leads I, II, and aVF (Pastore et al., 2016). It is well known that the normal cardiac electrical impulse begins in the sinusal node, and then spreads throughout the atrial chambers, defining the P-wave regular morphology. When the atrial depolarization wavefront spreads in an abnormal way, and/or presents accessory pathways, reentries, or conduction delays, the P-wave morphology may severely change (Dilaveris and Gialafos, 2002; Martínez et al., 2015). In this context, fragmented or longer P-waves have been associated with atrial conduction disturbances, sinus impulse blocks, increased left atrium, and hypertension (Platonov, 2012; Martínez et al., 2015). Recognizing that the extension of atrial time impulse conduction and the nonhomogeneous propagation of sinusal electrical impulse are indicative physiological features for the atria propense to atrial fibrillation, the morphological analysis of the P-wave is particularly important for stratifying risks concerning the predisposition of a given patient to suffer from atrial fibrillation, who may have benefits from effective preventive treatments (Platonov, 2012).

T-wave and P-wave detection and delineation are commonly taken as the third step within ECG signal segmentation and feature extraction, that is, just after QRS detection and delineation. As both T- and P-waves present considerably lower energy levels than QRS complexes, and that their morphology may assume several patterns, being easily masked by physiological noise and interference, a common practice is to establish search windows between already segmented QRS complexes for filtering and/or mathematical modeling to identify the occurrence (fiducial points) or absence and corresponding boundaries of those waveforms.

In the next session, we will describe and evaluate a group of different already published approaches for estimating the occurrence times or absence of P-waves and T-waves, as well as their corresponding boundaries under different scenarios: clean signals, noise interference, physiological noise, different arrhythmia classes, morphology alterations or disturbances, and different leads from a same record. In the following section, we will focus on the description of innovative approaches for mathematical modeling of P-waves and T-waves with the aim of both detecting and delineating those waveforms, and obtaining additional features, which can be extracted from the fitted models beyond the traditional time-domain parameters.

6.2 ANALYZING DIFFERENT APPROACHES FOR P/T-WAVE DETECTION AND DELINEATION

Two of the most cited works, Martínez et al. (2004) and Ghaffari et al. (2009), apply different scale factors within Discrete Wavelet Transform (DWT) for detecting and delineating P-waves and T-waves. Martínez et al. (2004), initially, defines a search window for each beat, considering a computed RR interval. Taking the resultant DWT over the search window, the approach searches for local maxima of $|W_{2^4}x[n]|$, that is, the modulus of the filtered signal associated with the scale factor 2^4. If at least two critical points of opposite signs exceed a threshold ε_T, a T-wave is considered to be present. For a given search window, the threshold ε_T is computed as

$$\varepsilon_T = 0.25\,RMS(W_{2^4}x[n]), \tag{6.1}$$

where the RMS operation is computed, considering the filtered version of the interval between two consecutive already delineated QRS complexes.

Considering that a T-wave is present, a pair of consecutive local maxima and local minima within the filtered signal, with amplitude greater than γ_T, are associated with significant slopes of the original waveform, and the zero crossings between them as T-wave peaks. According to the number, order, and polarity of the found critical points, the methodology identify one of six possible T-wave morphologies: positive (a positive critical point followed by a negative critical point), negative (a negative critical point followed by a positive critical point), biphasic (three critical points, $+/-/+$ or $-/+/-$), upward (a single one positive critical point exceeding the threshold γ_T), and downward (a single one negative critical point exceeding the threshold γ_T). For a given search window, the threshold γ_T is computed as

$$\gamma_T = 0.125\max(|W_{2^4}x[n]|), \tag{6.2}$$

where the operator maximum of the modulus is computed, considering the filtered version of the interval between two consecutive already delineated QRS complexes.

Figs. 6.1(A)–(D) illustrate four different T-wave morphologies and the corresponding Wavelet transform scaled versions (scale factor equal to 2^4), considering ECG records from MIT–BIH Arrhythmia Database (mitdb) and QT Database (qtdb). Each figure presents the computed thresholds ε_T (lower one) and γ_T (higher one) within the lower graphs. Concerning the first T-wave (Fig. 6.1(A)), record 101/mitdb, we observe a positive morphology within the ECG window, two significant critical points (firstly the positive one and thereafter the negative one) exceeding both thresholds ε_T and γ_T within the filtered signal, and the zero-crossing between them associated with T-wave peak. Fig. 6.1(B), record 118/mitdb, shows a negative T-wave, which is evidenced by two signs of opposed critical points exceeding both thresholds ε_T and γ_T, firstly the negative one and thereafter the positive one, with the zero-crossing between them associated with T-wave peak. Fig. 6.1(C) presents a biphasic T-wave $(+/-)$, record sele0704/qtdb, evidenced by three critical points $(+/-/+)$ and their corresponding two zero-crossings. Finally, Fig. 6.1(D) presents an upward

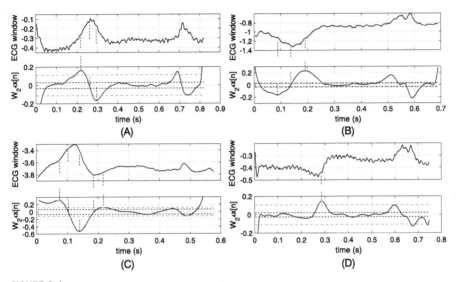

FIGURE 6.1

Applying Wavelet transforms and computed thresholds over T-wave search windows: estimating T-wave time occurrences and the corresponding morphologies. (A) Record 101/mitdb: positive T-wave. (B) Record 118/mitdb: negative T-wave. (C) Record sele0704/qtdb: biphasic T-wave ($+/-$). (D) Record 100/mitdb: upward T-wave.

T-wave with two sign-opposed critical points exceeding the lower threshold (ε_T), but just one critical point (the positive one) exceeding the higher threshold (γ_T).

Martínez et al. (2004) also considers to search for T-wave occurrence within the scaled version $W_{2^5}x[n]$ if the algorithm has not been able to detect significant critical points within $W_{2^4}x[n]$. The T-wave peak mapping is accomplished by searching within $W_{2^3}x[n]$ for zero-crossings between the already detected sign-opposed critical points. Aiming to identify T-wave edges, the approach applies other two thresholds, $\varepsilon_{T_{on}}$ and $\varepsilon_{T_{end}}$. Firstly, the algorithm identifies the first and the last critical points, n_{first} and n_{last}, respectively, associated with T-wave. Then the thresholds are computed as

$$\varepsilon_{T_{on}} = 0.25|W_{2^4}x[n_{first}]|, \tag{6.3}$$

$$\varepsilon_{T_{end}} = 0.4|W_{2^4}x[n_{last}]|. \tag{6.4}$$

Then candidates to T-wave onset are determined by applying the follow criteria: (i) search for the sample before n_{first}, where $|W_{2^4}x[n]|$ is below $\varepsilon_{T_{on}}$; (ii) search for a local minimum of $|W_{2^4}x[n]|$ before n_{first}. The selected location for T-wave onset is the candidate located at the nearest distance from the zero-crossing associated with T-wave peak. Analogously, candidates for T-wave end are determined by applying the follow criteria: (i) search for the sample after n_{last}, where $|W_{2^4}x[n]|$ is below $\varepsilon_{T_{end}}$; (ii) search for a local minimum of $|W_{2^4}x[n]|$ after n_{last}. The selected location

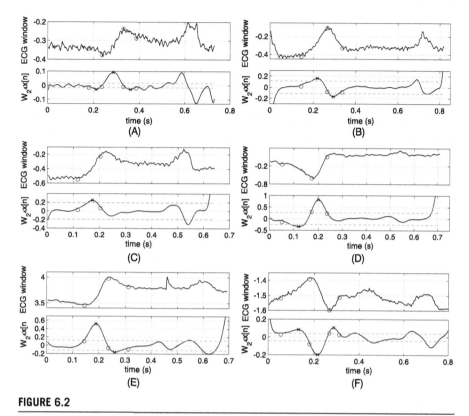

FIGURE 6.2

Segmentation of T-waves presenting different morphologies. (A) Record 100/mitdb: T-wave segmented and classified as biphasic (−/+). (B) Record 101/mitdb: T-wave segmented and classified as positive. (C) Record 105/mitdb: T-wave segmented and classified as upward. (D) Record sel14157/qtdb: T-wave segmented and classified as negative. (E) Record sele0303/qtdb: T-wave segmented and classified as positive. (F) Record sele0126/qtdb: T-wave segmented and classified as biphasic (+/−).

for T-wave end is the candidate located at the nearest distance from the zero-crossing associated with T-wave peak.

Illustrating aspects of the approach proposed by Martínez et al. (2004), Figs. 6.2(A)–(F) present the processes of T-wave detection and segmentation (peaks, onsets, and ends), and also morphology, classifying within the records 100/mitdb, 101/mitdb, 105/mitdb, sel14157/qtdb, sele0303/qtdb, and sele0126.

Concerning the obtained results, Fig. 6.2(A) presents a biphasic T-wave (−/+), recognized through the identification of three critical points (−/ + /−) exceeding threshold γ_T and two zero-crossings between them. Figs. 6.2(B) and 6.2(E) present cases of positive T-waves associated with the identification of two critical points (+/−) exceeding the referred threshold, with a zero-crossing between them. Fig. 6.2(C) presents a T-wave, classified as an upward waveform, due to the identification of a single critical point (+) exceeding γ_T. In this case, the algorithm has only identified T-wave onset and the offset. Fig. 6.2(D) presents a negative T-wave, associated with the presence of two critical points $(-/+)$ exceeding γ_T, with a zero-crossing between them. And finally, Fig. 6.2(F) presents a biphasic T-wave $(+/-)$, associated with the identification of three critical points $(+/-/+)$ exceeding γ_T and two zero-crossings between them.

For detecting and segmenting P-wave, the corresponding process is analogous. However, a specific search window should be carefully delimited, having in mind that P-wave amplitudes are commonly the lowest ones, and, for some pathologic conditions, P-wave is absent. Despite the authors not mentioning how they selected the RR-dependent interval for investigating the occurrence of P-wave and delineating the corresponding waveform, a satisfactory suggestion is to take a search window from each T-wave end to the subsequent QRS onset.

Once we have delimited such window, we compute the following parameters

$$\varepsilon_P = 0.02\,RMS(W_{2^4}x[n]), \tag{6.5}$$

where the RMS metric is computed within the whole interval between two consecutive QRS end and QRS onset edges,

$$\gamma_P = 0.125\max(|W_{2^4}x[n]|), \tag{6.6}$$

where the time index n pertains to the delimited P-wave search window.

Therefore a P-wave is considered to be present if at least two critical points of opposite signs exceed the threshold ε_P within the specific P-wave search window. Furthermore, once we have checked the occurrence of P-wave, the critical points with amplitude exceeding the threshold γ_P are associated with significant slopes of the original waveform, and the corresponding zero-crossings with P-wave peaks.

According to the number, order, and polarity of the found critical points exceeding γ_P, the methodology identifies the following P-wave morphologies: positive (a positive critical point followed by a negative critical point), negative (a negative critical point followed by a positive critical point), and biphasic (three critical points, $+/-/+$ or $-/+/-$). To identify the P-wave limits, the approach applied exactly the same criteria used for delineating T-wave, but considering the thresholds

$$\varepsilon_{P_{on}} = 0.5|W_{2^4}x[n_{first}]|, \tag{6.7}$$

$$\varepsilon_{P_{end}} = 0.9|W_{2^4}x[n_{last}]|, \tag{6.8}$$

where n_{first} and n_{last} are, respectively, the first and the last critical points (whose amplitudes exceed γ_P) associated with P-wave.

Figs. 6.3(A)–(E) present the processes of P-wave detection and segmentation (peaks, onsets, and ends), and also morphology, classifying within the records 100/mitdb (channel 1), 101/mitdb (channel 1), 108/mitdb (channel 2), sel123/qtdb (channel 2), and 103/mitdb (channel 2).

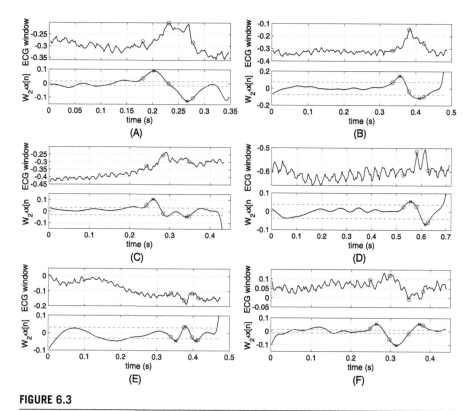

FIGURE 6.3

Segmentation of P-waves presenting different morphologies. (A) Record 100/mitdb (channel 1): P-wave segmented and classified as positive. (B) Record 101/mitdb (channel 1): P-wave segmented and classified as positive. (C) Record 108/mitdb (channel 2): P-wave segmented and classified as positive. (D) Record sel123/qtdb (channel 2): P-wave segmented and classified as positive. (E) Record 103/mitdb (channel 2): P-wave segmented and classified as biphasic ($-/+$). (F) Record 111/mitdb (channel 2): P-wave segmented and classified as biphasic ($+/-$).

All the P-wave tracings illustrated by the Figs. 6.3(A)–(F) present a common characteristic: their amplitudes and morphologies are clearly masked by noise. Figs. 6.3(A)–(D) present all of them, cases of positive morphologies, with energy, duration, and geometry of the corresponding waveforms being significantly different. The corresponding Wavelet transforms present pairs of sign-opposed critical points exceeding the threshold γ_P. Fig. 6.3(C) shows that some protection measures should be applied, mainly related to isolated critical points and/or consecutive critical points with the same sign. Figs. 6.3(E) and 6.3(F) show episodes of biphasic P-waves: ($-/+$) and ($+/-$). They are recognized by the presence of three consecutive sign-opposed critical points exceeding the threshold γ_P. Here we emphasize that a maximum number of three critical points is related to a given P-wave morphology.

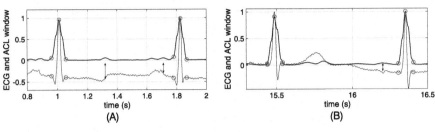

FIGURE 6.4

Definition of search windows and association between local maxima within the ACL signal and P/T-waves occurrence within ECG search windows. (A) Record 100/mitdb: T-wave and P-wave detection through the identification of local maxima within ACL signal. (B) Record 103/mitdb: T-wave and P-wave detection through the identification of local maxima within ACL signal.

Ghaffari et al. (2009) obtained five scaled versions related to the Discrete Wavelet Transform of the ECG input signal: scales 2^1, 2^2, 2^3, 2^4, and 2^5. After detecting and delineating each QRS complex using the area-curve length (ACL) signal, computed within scale 2^3 or 2^4, the proposed approach allows the detection and delineation of P- and T-waves. For this, a specific scaled version is taken, for example, the scale 2^3, and a new ACL signal is derived. Then a local search for two local maxima is performed within the ACL signal between the locations related to QRS offset and the subsequent QRS onset. Actually, the search interval is divided into two segments each corresponding to approximately half the RR interval. It is expected that at least one dominant maximum be found in each of these intervals: the local maximum close to the left QRS offset is associated with a T-wave peak, and the maximum close to the right QRS onset is associated with a P-wave peak.

Figs. 6.4(A), (B) illustrate the association between the local maxima within the ACL signal (computed over wavelet transform, scale 2^3) and the T-wave and P-wave occurrences within the interval correspondent to the segment between a given QRS onset and the subsequent QRS offset. The signals are excerpts extracted from the records 100/mitdb (channel 1) and 103/mitdb (channel 2).

Concerning each waveform, other specific search windows are established for identifying the wave edges. Thus the algorithm performs a search for local minima between the preceding QRS onset and T-wave peak for delineating the beginning of the T-wave. Analogous searches are performed between the T-wave peak and half of the RR-interval for delineating T-wave, between the half of the RR-interval and the P-wave peak for delineating the onset of a P-wave, and, finally, between the P-wave peak and the onset of the subsequent QRS complex for delineating the end of a P-wave. Alternatively, the algorithm searches for the position, where ACL slope is less than 1/15–1/20 of the maximum slope computed for the respective window and associates that sample with a wave edge.

Figs. 6.5(A), (B) present a schema for detecting and segmenting T- and P-waves within a RR-dependent interval. Thus in both situations, T-wave is mapped as a local

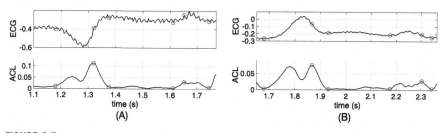

FIGURE 6.5

Process of T-wave and P-wave segmentation: use of ACL signal for mapping samples with maximum and minimum energy within multiple search windows. (A) Record 100/mitdb:T-wave and P-wave segmented through the identification of local maxima and samples with minimum energy. (B) Record 103/mitdb:T-wave and P-wave segmented through the identification of local maxima and samples with minimum energy.

maximum within the search interval beginning at previous QRS onset and ending at half the RR-interval. P-wave is also mapped as a local maximum, but within the interval beginning at half the corresponding RR-interval and ending at subsequent QRS offset. The wave edges are mapped in the ACL signal as samples with local minimum energy at their corresponding search intervals. Some metrics protecting against bad detections should be taken as minimum/maximum wave lengths and minimum power for P-wave, having in mind that, for some situations, P-wave is absent.

6.3 MATHEMATICAL MODELING OF P-WAVES AND T-WAVES

A family of methodologies recently published is concerned with finding and testing several mathematical models for fitting T-wave and P-wave patterns. The goal is not only to detect and delineate the waveforms within an accurate approach, but also to capture some intrinsic information related to different morphologies such as width, asymmetry, small amplitude variations, fragmentation, eventual distortions or subtle changes.

In this sense, Madeiro et al. (2013) proposed a mathematical model of a skewed Gaussian function applied for modeling T-wave, T-wave detection, and T-wave end location. Once QRS detection and delineation data are obtained, the approach establishes search windows between each QRS offset (J-point) and subsequent QRS onset, named $W[n]$. A zero-phase band-pass FIR filtering, using a Hamming window, is applied over each T-wave search window, considering that no overall preprocessing has been performed previously. The cutoff frequencies were selected as 0.01–12 Hz, based on experimental observations about T-wave spectral content, aiming to obtain noise attenuation while preserving physiological spectral content of the T-wave. The order of the filter is experimentally selected as 32. Then the approach proposes a training stage. Here the algorithm searches for the most appropriated parameters for a synthesized mathematical model (a skewed Gaussian function), which allows for a

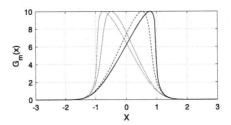

FIGURE 6.6

Four possible realizations for the derived skewed Gaussian function, corresponding to the α values $\pi/40$, $\pi/60$, $-\pi/40$, and $-\pi/60$.

satisfactory modeling of T-wave waveform. These parameters are the standard deviation of a Gaussian function (σ) and a distortion angle (α).

Before the explanation concerning the strategy for finding the optimal parameters, we describe the proposed mathematical model: skewed Gaussian function. Thus given a Gaussian function $G(x)$ with a maximum amplitude equal to A_G and standard deviation σ_1, defined within the interval $x_u \leq x \leq x_l$, and given a linear function $f(x)$ with a slope equal to α and a null y-intercept parameter, the algorithm derives a resulting function $G_m(x)$, such that

$$G(x) = A_G \frac{1}{\sqrt{2\pi}\sigma_1} e^{\frac{-x^2}{2\sigma_1^2}},$$ (6.9)

$$G_m(x) = G(x - \rho(x)),$$ (6.10)

$$\rho(x) = f(G(x)) = \tan(\alpha)G(x).$$ (6.11)

As we can see, depending on the relative amplitude of $G(x)$, the original function is pulled to the right or to the left (according to the α value) in a greater or lesser extent. We may say that the distortion is a linear function varying with the Gaussian function amplitude, and also it is linear dependent on the angular coefficient α. Thus the technique produces a distorted and asymmetric synthetic signal $G_m(x)$ from the Gaussian function $G(x)$. Fig. 6.6 illustrates four possible realizations of the proposed synthetic signal $G_m(x)$ for α values equal to $\pi/40$, $\pi/60$, $-\pi/40$, and $-\pi/60$.

The approach developed by Madeiro et al. (2012) initially needs to learn the predominant pattern related to the T-waves of a given ECG record. Thus the authors established a training stage. For each one of the first N RR-dependent intervals (i.e. intervals between a given QRS offset and the subsequent QRS onset), one applies the cross-correlation function between the corresponding search window $W[n]$ and a family of skewed Gaussian functions (patterns $G_m(x)$) related to a specific group of σ and α values. The set of values for σ and α may be adjusted according to the sampling frequency of the processed ECG signals. Considering, for example, the records from QT Database (QTDB, qtbd), for which the sampling frequency is 256 Hz, a satisfactory choice is $\sigma \in [0.2, 1.5]$ and $\alpha \in [-\pi/20, \pi/20]$.

Then if we denote each parameterized skewed Gaussian function as $G_m^{\sigma,\alpha}[n]$, the signal resultant from the cross-correlation operation is given as

$$S[n] = \sum_{\tau=0}^{K} W[\tau] G_m^{\sigma,\alpha}[n - \tau]. \tag{6.12}$$

The location of the maximum value of $|S[n]|$ is identified and denoted as $Pk_{\alpha\sigma}^{(1)}$. Then we established a new window around $Pk_{\alpha\sigma}^{(1)}$, denoted $W_S[k]$, and computed an error function $E_{\sigma,\alpha}^{(1)}$, given as

$$E_{\sigma,\alpha}^{(1)} = \frac{\sum_{k=Pk_{\alpha\sigma}^{(1)}-\gamma_1}^{Pk_{\alpha\sigma}^{(1)}+\gamma_2} |W_S[k] - \hat{G}_m^{\sigma,\alpha}[k]|^2}{\sum_{k=Pk_{\alpha\sigma}^{(1)}-\gamma_1}^{Pk_{\alpha\sigma}^{(1)}+\gamma_2} |W_S[k]|^2}, \tag{6.13}$$

where γ_1 and γ_2 are offset times, which delimit the window $W_S[k]$, and $\hat{G}_m^{\sigma,\alpha}[k]$ is a normalized version of $G_m^{\sigma,\alpha}[k]$ according to the maximum and minimum values of $|W_S[k]|$, and aligned by its peak amplitude location, coinciding with the relative location of $Pk_{\alpha\sigma}^{(1)}$ within $W_S[k]$.

By computing $E_{\sigma,\alpha}^{(1)}$ for all possible members of the ordained pair (σ,α), we searched and found a candidate optimal pair (σ_s,α_s) for each search window $W[n]$. Then for each $W[n]$ and given the specific estimated T-wave peak location $Pk_{\alpha\sigma}^{(1)}$ found with (σ_s,α_s), here denoted as $Pk_{\alpha_s\sigma_s}^{(1)}$, we again computed an evaluation error function $E_{\sigma,\alpha}^{(2)}$ for each ordained pair (σ,α), given as

$$E_{\sigma,\alpha}^{(2)} = \frac{\sum_{k=Pk_{\alpha_s\sigma_s}^{(1)}-\gamma_1}^{Pk_{\alpha_s\sigma_s}^{(1)}+\gamma_2} |W_S[k] - \hat{G}_m^{\sigma,\alpha}[k]|^2}{\sum_{k=Pk_{\alpha_s\sigma_s}^{(1)}-\gamma_1}^{Pk_{\alpha_s\sigma_s}^{(1)}+\gamma_2} |W_S[k]|^2}. \tag{6.14}$$

This process is required because each individual combination (σ,α) produces an independent location for the maximum value of $|S[n]|$ (Eq. (6.12)) and therefore for T-wave peak estimation.

Finally, the optimal parameters related to the skewed Gaussian function, which best fits a given T-wave, denoted as (σ_o,α_o), are searched for by minimizing the function $E_{\sigma_s,\alpha_s}^{(1)}$ for each search window $W[n]$. The approach averages the individual values of pairs (σ_o,α_o) found for each of N search windows $W[n]$ (defined for a training stage), and obtain the parameters $(\overline{\sigma}_o,\overline{\alpha}_o)$. The minimum values found for $E_{\sigma,\alpha}^{(2)}$ are also averaged and the resultant mean value is denoted as $E_{\min}^{(2)}$.

We may say that the optimal parameters ($\overline{\sigma}_o$ and $\overline{\alpha}_o$) parameterize the proposed skewed Gaussian function, which characterize a given prevailing T-wave morphology. Then the computing required for estimating each T-wave peak location is performed by applying the cross-correlation function between each search window $W[n]$

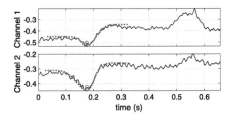

FIGURE 6.7

Examples of different T-wave morphologies (continuous line) and the matched skewed Gaussian functions (dashed line) for each waveform.

and the optimal skewed Gaussian function

$$S[n] = \sum_{\tau=0}^{K} W[\tau] G_m^{\overline{\sigma_o},\overline{\alpha_o}}[n - \tau].$$ (6.15)

In order to make the algorithm efficient, the proposed approach is not applied in searching for optimal parameters (σ_o, α_o) within each RR-dependent interval, which would demand significant computing time. Instead the same error function given by Eq. (6.13) is computed for each search window and the normalized skewed Gaussian function $\hat{G}_m^{\sigma_o,\alpha_o}[n]$. If for a given T-wave detection, the error function $E_{\sigma_o,\alpha_o}^{(1)}$ is found higher than $\theta E_{min}^{(2)}$, where θ is a tolerance factor, then the algorithm performs another learning process for that specific search window.

Fig. 6.7 illustrates the T-wave modeling by the approach proposed by Madeiro et al. (2012) within both available leads from the record sel100/qtdb (QT Database). The algorithm allows finding the optimal parameters $(\sigma_o, \alpha_o) = (0.48; -0.0314 \text{ rad})$, for the first channel, and $(\sigma_o, \alpha_o) = (0.58; 0.0209 \text{ rad})$ for the second channel.

Concerning the occurrence of biphasic T-waves, the proposed approach monitors within each filtered signal $S[n]$ (resultant from the auto-correlation operation), the absolute values exceeding 70% of the maximum value of $|S[n]|$. If the algorithm finds a local maximum and a local minimum, the sign of the obtained optimal skewed Gaussian function is checked. Then the critical point chosen as T-wave peak is that one with the same amplitude sign as the one related to the optimal skewed Gaussian function.

Beyond the T-wave modeling and T-wave peak estimation, the algorithm developed by Madeiro et al. (2012) also allows the detection of each T-wave end. For this task, initially, the method needs to estimate if a given T-wave morphology is monophasic or biphasic. To do this, for each sample within each search window $W[n]$, the following moving averages are computed (Zhang et al., 2006) as follows:

$$\overline{s_k} = \frac{1}{2p + 1} \sum_{j=k-p}^{k+p} W[j],$$ (6.16)

$$A_k = \sum_{j=k-w+1}^{k} (W[j] - \overline{s_k}), \tag{6.17}$$

where p is the number of samples corresponding to 16 ms, and w is the number of samples corresponding to 128 ms. Then we identify the location of maximum (k_1) and minimum (k_2) critical points for the signal A_k inside its corresponding search window $W[n]$. For a given threshold λ, one tests the condition (Zhang et al., 2006)

$$\frac{1}{\lambda} < \frac{|A_{k_1}|}{|A_{k_2}|} < \lambda. \tag{6.18}$$

If this condition is true for a given parameter λ, then the corresponding T-wave is classified as biphasic. Otherwise, it is classified as monophasic. If a monophasic T-wave is recognized, then the following sequence of tasks is performed:

- The position of the maximum slope of the optimal skewed Gaussian function, after the peak location, is determined. We denote the detected position as T_{der} and the corresponding slope value as M_{der};
- The location of the first sample after T_{der} within the optimal skewed Gaussian function for which the slope is below $\beta_1 M_{der}$, where β_1 is a percentage factor, is determined. We denote the corresponding position as T_{min};
- Then the locations M_{der} and T_{min} are mapped into the search window $W[n]$ by aligning the peak of the optimal skewed Gaussian function with the estimated T-wave peak location.

After mapping the reference locations M_{der} and T_{min}, the approach applies the Trapezium area technique, proposed by Vázquez-Seisdedos et al. (2011) for estimating T-wave end. Given a moving sample location T_e, where $T_{der} \leq T_e \leq T_{min}$, the vertices of a trapezium are defined for each value of location T_e by the following ordained pairs (x, y): $(T_{der}, W[T_{der}])$, $(T_e, W[T_e])$, $(T_{min}, W[T_e])$, and $(T_{min}, W[T_{der}])$. The area of the trapezium, which is a function of the location T_e, whose dimensions are computed according to the referred ordained pairs, is given as

$$A_T(T_e) = 0.5(W[T_{der}] - W[T_e])(2T_{min} - T_e - T_{der}). \tag{6.19}$$

Then we search for the maximization of $A_T(T_e)$ by varying T_e location, and the T-wave end is associated with the position T_e that allows for the maximization of $A_T(T_e)$.

If a biphasic T-wave is recognized, the approach seeks to recognize if the modeled T-wave phase is the first or the second one. Firstly, we check if k_1 (maximum critical value of A_k) is located before or after k_2 (minimum critical value of A_k) (see Eqs. (6.16)–(6.18)). Then we check if a given optimal skewed Gaussian function is positive or negative. Therefore if the fitted mathematical model is associated with the second phase of a biphasic T-wave, the process for estimating T-wave end follows the same process detailed above, for a monophasic T-wave. Otherwise, if we conclude that the kernel models the first phase, some specific tasks are required:

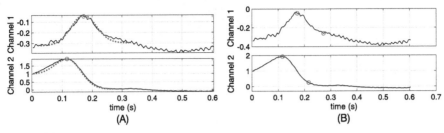

FIGURE 6.8

Process of T-wave modeling and segmentation: using reference points of maximum and minimum slope for performing trapezium-area approach. (A) Record sel102/qtdb: modeled T-waves as positive skewed Gaussian functions, considering the two available leads. (B) Record sel102/qtdb: Detection of T-wave peak and end through the trapezium-area approach.

- The approach establishes another window which should contain the peak and the end of the second phase: $W_2[n]$. The beginning of $W_2[n]$ is defined as the position T_{der}, identified within the skewed Gaussian function and mapped within the first T-wave phase, and the end is the subsequent QRS offset;
- The critical point related to the peak of the second phase is searched between the positions k_1 and k_2. The specific position is denoted T_P^2;
- We search for a local maximum of the slope of $W_2[n]$ after T_P^2, not exceeding a given maximum distance from T_P^2. The corresponding position and the slope value are denoted $T_{der}^{(2)}$ and $M_{der}^{(2)}$;
- As we did for monophasic T-waves, we identify the first sample location, where the slope value is below $\beta_2 M_{der}^{(2)}$, where β_2 is a percentage or tolerance factor. The corresponding position is named $T_{min}^{(2)}$, and it should not exceed a maximum distance from T_P^2, for example, 100 ms;
- For a given moving sample location $T_e^{(2)}$, obeying $T_{der}^{(2)} \leq T_e^{(2)} \leq T_{min}^{(2)}$, one computes the trapezium area expression $A(T_e^{(2)})$. The T-wave end is located by searching the value of $T_e^{(2)}$, which minimizes $A(T_e^{(2)})$.

Illustration the whole proposed approach, Figs. 6.8(A), (B) present specific cases of positive T-wave morphologies within record sel102/qtdb, with their corresponding best-fitted kernels and respective points of segmentation: T-wave peak and T-wave end.

As a continuation and deepening of the research carried out through the work published in Madeiro et al. (2012), Madeiro et al. (2017) applied, evaluated, and compared three different mathematical kernels aiming to model and segment both T-waves and P-waves. The functions are: a simple Gaussian function, a composition of two Gaussian functions, and the Rayleigh probability density function.

As we have already detailed in the previous chapter, within the topic related to QRS modeling, a single and normalized discrete-time Gaussian function may be defined as (Madeiro et al., 2017)

$$G_{\sigma_1}[k] = \exp(\frac{k^2}{2\sigma_1^2}),$$ (6.20)

where k may vary within a given interval $-x_1 \leq k \leq x_2$, and the parameter σ_1 is related to the Gaussian width. Madeiro et al. (2017) consider the interval $-3 \leq k \leq 3$.

The proposed mathematical model resulting from combining two discrete-time Gaussian functions with individual width parameters σ_1 and σ_2 is given as

$$G_{\sigma_1,\sigma_2}[k] = \begin{cases} G_{\sigma_1}[k] & \text{if } -3 \leq k \leq 0, \\ G_{\sigma_2}[k] & \text{if } 0 \leq k \leq 3. \end{cases}$$ (6.21)

Finally, the discrete-time Rayleigh probability density function $R_{\sigma_1}[k]$ is defined within the interval $0 \leq k \leq 10$:

$$R_{\sigma_1}[k] = \frac{k}{\sigma_1^2} \exp(-\frac{k^2}{2\sigma_1^2}),$$ (6.22)

where σ_1 is the corresponding variable parameter.

Additionally, the authors have defined a modification of the original Rayleigh probability density function $R_{\sigma_1}^m$, also comprising the interval $0 \leq k \leq 10$, given as its time-reversal

$$R_{\sigma_1}^m[k] = R_{\sigma_1}[10 - k].$$ (6.23)

Based on the foregoing modeling of each T-wave and P-wave, Madeiro et al. (2017) performs T-wave and P-wave peak detection by adapting the approach developed by Martínez et al. (2004). Thus considering that each QRS complex is accurately segmented, the approach delimits search windows between each QRS complex end and subsequent QRS complex offset. After applying the DWT over the referred to search windows, using scale factor 2^4 and a quadratic spline as wavelet-mother (Martínez et al., 2004), the algorithm searches for two local maxima within the filtered signal $|W_{2^4}x[n]|$. If at least two sign-opposed critical points are identified exceeding a threshold ε_T, computed as $0.25 RMS(W_{2^4}x[n])$, the time interval between them does not exceeding 200 ms, and the location of the first critical point does not exceed 65% of the search interval length, a T-wave is considered to be present. In the case of more than two critical points satisfying such conditions, we take the pair of critical points, whose product of the corresponding amplitudes is the highest among the tested pairs. For a given pair of identified critical points, we also check if their amplitudes exceed a threshold γ_T, computed as $0.125 \max(|W_{2^4}x[n]|)$. If this condition is validated for both critical points, then they are mapped as significant slopes within the original search window, and the zero-crossing located at the intermediary region is associated with T-wave peak. If just one critical point exceeds the threshold γ_T, we conclude that the detected waveform presents a single significant slope (upward or downward). In this situation, we take as reference for T-wave peak location the maximum or minimum critical point with the highest amplitude within the original signal located between the first zero-crossing within the filtered signal, before the detected significant slope, and the first zero-crossing after the detected significant slope within the filtered signal.

Once a given T-wave peak location is estimated, the approach performs the process of T-wave modeling. Each one of the proposed kernels presents one or two specific parameters, corresponding to the standard deviations σ_1 and σ_2. We established a process of matching each kernel with each waveform, by varying the parameters over a predefined range: $\sigma_i \in [0.2; 1.5]$ with a time-resolution equal to 80 samples.

Concerning the kernel $G_{\sigma_1}[k]$, we established a search window $W_1[n]$ around each T-wave peak, starting 150 ms before the corresponding fiducial point and finishing 150 ms after the peak. Then we applied a process for normalizing the kernel samples, such that: its maximum amplitude would be equal to the maximum amplitude of $|W_1[n]|$, its minimum amplitude observed before its maximum critical point would be equal to the first minimum of $|W_1[n]|$ (the lowest amplitude observed from the starting point to the maximum critical point), its minimum amplitude observed after its maximum critical point would be equal to the second minimum of $|W_1[n]|$ (the lowest amplitude observed from the maximum critical point to the last sample).

Then the normalized kernel $G_{\sigma_1}^m[k]$ is aligned by its peak location with T-wave peak within the search window $W_1[n]$. Finally, we computed a normalized RMS error $E_{\sigma_1}^{(1)}$ for each tested parameter σ_1 and each analyzed waveform, given as

$$E_{\sigma_1}^{(1)} = \sqrt{\frac{\sum_{k=Pk-\gamma}^{Pk+\gamma} |W_1[k] - G_{\sigma_1}^m[k]|^2}{\sum_{k=Pk-\gamma}^{Pk+\gamma} |W_1[k]|^2}}, \tag{6.24}$$

where Pk refers to the position of the T-wave peak inside the search window $W_1[n]$, and γ refers the number of samples corresponding to 150 ms.

Therefore we performed an automatic search for the value of σ_1, which allows the computing of the minimum value of $E_{\sigma_1}^{(1)}$, named as $E_{\min}^{(1)}$, which is related to the normalized RMS error between the optimal kernel (best fitted) and a given waveform.

Analogously, we proceeded with the steps for normalizing and peak alignment for the other kernels G_{σ_1,σ_2} (composition of Gaussian functions), $R_{\sigma_1}[k]$ (Rayleigh probability density function), and $R_{\sigma_1}^m$ (modified Rayleigh probability density function). The computing of the respective evaluation parameters $E_{\sigma_1,\sigma_2}^{(2)}$, $E_{\sigma_1}^{(3)}$, and $E_{\sigma_1}^{(4)}$, and the search for the optimal kernels are also analogous processes. For the kernel G_{σ_1,σ_2}, all the possible combinations of σ_1, σ_2, within a given range, should be evaluated.

Figs. 6.9 illustrate the results for the modeling of T-wave morphologies within records from QT Database using the four detailed mathematical functions: Gaussian function (6.9(A)), composition of two Gaussian functions (6.9(B)), the Rayleigh probability density function (6.9(C)), and the time-reversal of the Rayleigh probability density function (6.9(D)). It is clear that the composition of two Gaussian functions allows for a higher flexibility to model different T-wave morphologies than

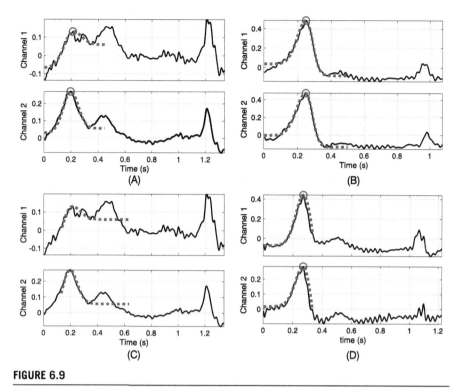

FIGURE 6.9

Different approaches of T-wave modeling using: (A, sel14172/qtdb) Gaussian function, (B, sel117/qtdb) composition of two Gaussian functions, (C, sel14172/qtdb) Rayleigh probability density function, and (D, sel123/qtdb) the time-reversal of the probability density function.

that related to a simple Gaussian function. However, the use of Rayleigh probability function (an intrinsically asymmetrical function) provides a significant robustness for modeling, using a single variable parameter for tuning.

Concerning the delineation of T-wave, the methodology for identifying the samples related to the onset and offset for a given waveform is based on the locations of strategic sample points specifically related to each one of the mathematical models (Madeiro et al., 2013; Madeiro et al., 2017). Thus we executed the following sequence of tasks to identify T-wave end:

- Given a fitted mathematical function, we searched for the maximum absolute slope, among the samples located after the maximum critical point (peak of the kernel). We denoted the sample related to the maximum slope and its corresponding value, respectively, as T_{der} and M_{der};
- Among the samples located after T_{der}, we searched for the first sample whose slope absolute value was below 5% of the M_{der} value. We denoted the sample found with the referred to restrictions as T_{min};

- The samples T_{der} and $T_{\min}$ were mapped within the search window $W_1[n]$, and then we applied the trapezium area methodology proposed by Vázquez-Seisdedos et al. (2011) for estimating T-wave end.

Thus given a moving sample location T_e, where $T_{\text{der}} \leq T_e \leq T_{\min}$, the vertices of a trapezium were defined for each value of location T_e by the following ordained pairs (x, y): $(T_{\text{der}}, W_1[T_{\text{der}}])$, $(T_e, W_1[T_e])$, $(T_{\min}, W_1[T_e])$, and $(T_{\min}, W_1[T_{\text{der}}])$. The area of the trapezium, considering the foregoing vertices, as already detailed in the previous section, was given as

$$A_T(T_e) = 0.5|W[T_{\text{der}}] - W[T_e]|(2T_{\min} - T_e - T_{\text{der}}). \tag{6.25}$$

Then we searched for the maximization of $A_T(T_e)$ by varying T_e location, and the T-wave end was associated with the position T_e that allowed for the maximization of $A_T(T_e)$ (Vázquez-Seisdedos et al., 2011; Madeiro et al., 2012; Madeiro et al., 2017).

For detecting T-wave onset, one executes an analogous sequence of tasks, taking as reference the signal segment within each mathematical model and within a given search window $W_1[n]$ located before their corresponding maximum critical points:

- Given a fitted mathematical function, we searched for the maximum absolute slope, among the samples located before the maximum critical point (peak of the kernel). We denoted the sample related to the maximum slope and its corresponding value respectively as T_{der} and M_{der};
- Among the samples located before T_{der}, we searched for the last sample, whose slope absolute value was below 5% of the M_{der} value. We denoted the sample found with the referred to restrictions as $T_{\min}$;
- The samples T_{der} and $T_{\min}$ were mapped within the search window $W_1[n]$, and then we applied the trapezium area methodology proposed by Vázquez-Seisdedos et al. (2011) for estimating T-wave end.

Again, we defined a moving sample location T_b, where $T_{\min} \leq T_b \leq T_{\text{der}}$. The vertices of a trapezium are defined, for each value of location T_e, by the following ordained pairs (x, y): $(T_{\text{der}}, W_1[T_{\text{der}}])$, $(T_b, W_1[T_b])$, $(T_{\min}, W_1[T_b])$, and $(T_{\min}, W_1[T_{\text{der}}])$. Considering the defined vertices, the area of the trapezium is given as

$$A_T(T_b) = 0.5|W[T_{\text{der}}] - W[T_b]|(T_{\text{der}} + T_b - 2T_{\min}). \tag{6.26}$$

The sample identified as the onset of an analyzed T-wave is the one whose value of $A_T(T_b)$ is maximum among all the computed values for $T_{\min} \leq T_b \leq T_{\text{der}}$.

Figs. 6.10(A)–(D) facilitate comparing and evaluating the T-wave segmentation results within a specific search window inside the record sel14157/qtdb (QT Database), based on strategic sample points mapped from the mathematical models: Gaussian function (Fig. 6.10(A)), composition of two Gaussian functions (Fig. 6.10(B)), Rayleigh probability density function (Fig. 6.10(C)), and the time-reversal of the Rayleigh function (Fig. 6.10(D)).

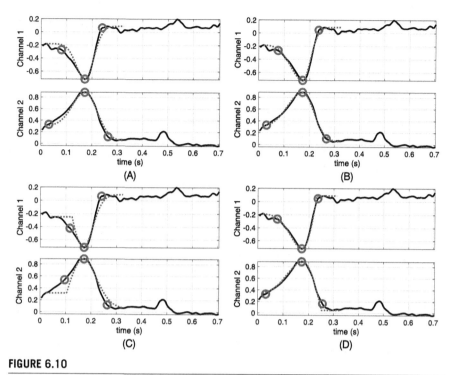

FIGURE 6.10

T-wave segmentation, considering the record sel14157/qtdb, based on strategic sample points mapped from the mathematical models: (A) Gaussian function, (B) composition of two Gaussian functions, (C) Rayleigh probability density function, and (D) the time-reversal of the Rayleigh function.

After the processes of T-wave peak detection, T-wave modeling and T-wave delineation, a new search window was established for each P-wave, corresponding to an interval between a given T-wave end and the subsequent QRS onset. As we did for T-wave peak estimation, we looked for two local maxima within $|W_{24}x[n]|$. If at least two signal-opposed critical points exceeded a threshold ε_P, computed as $0.02RMS(W_{24}x[n])$, the time interval between them not exceeding 200 ms, and the location of the last one not occurring after 30 ms to the end of the search window, a P-wave is considered to be present. If more than one pair of critical points satisfied the criteria, we selected the pair, whose product of amplitudes was the highest. Then the selected critical points were mapped as significant slopes within the search window and the zero-crossing located at the intermediary segment was mapped as the P-wave peak.

The processes for P-wave modeling and delineation are analogous to the respective tasks related to T-wave. We established an analysis window $W_2[n]$ around each P-wave peak, starting 100 ms before the fiducial point, and ending 100 ms after the peak. For each one of the proposed kernels, we performed a process of matching by

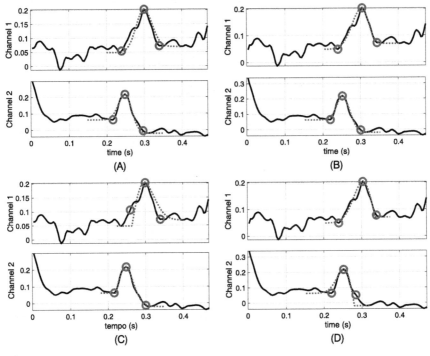

FIGURE 6.11

P-wave segmentation, considering the record sel14157/qtdb, based on strategic sample points mapped from the mathematical models: (A) Gaussian function, (B) composition of two Gaussian functions, (C) Rayleigh probability density function and (D) the time-reversal of the Rayleigh function.

varying the tuning parameters σ_1 and/or σ_2 within a predefined range $\sigma_i \in [0.2; 1.5]$, $i = 1, 2$, considering a time resolution of 60 samples. The specific task for normalizing kernel samples, a double-stage algorithm, was performed identically as we detailed for T-wave, and also the alignment between the peak of the kernel and the peak of the P-wave within the window $W_2[n]$. The computing of the normalized RMS error was also performed using Expression (6.24), where P_k refers to the relative position of the P-wave peak within the search window $W_2[n]$ and the parameter γ refers to the number of samples corresponding to 100 ms.

Finally the methodology for locating P-wave onset and offset, as for T-wave, was based on strategic sample points from each one of the tested mathematical models. Thus the sequence of tasks applied for identifying each P-wave onset and offset, and the equations for maximizing trapezium are identical to the ones applied for T-wave delineation (Eqs. (6.25) and (6.26)).

Figs. 6.11(A)–(D) facilitate comparing and evaluating the P-wave segmentation results within a specific search window inside the record sel14157/qtdb (QT Database), based on strategic sample points mapped from the mathematical models: Gaussian function (Fig. 6.11(A)), composition of two Gaussian functions (Fig. 6.11(B)), Rayleigh probability density function (Fig. 6.11(C)), and the time-reversal of the Rayleigh function (Fig. 6.11(D)).

6.4 EVALUATING AND COMPARING PERFORMANCES OF T-WAVE AND P-WAVE DELINEATORS

The results for T-wave and P-wave detection and delineation are obtained in terms of detection rates and time differences between automatic estimations for wave peaks and edges and manual annotations made by experts. In general, concerning detection rates, we may establish a time distance tolerance between the estimated wave peak location, indicated by an algorithm, and the reference manual annotated peak. If a given estimated peak is located in a sample exceeding a maximum time distance from the reference point (for example 120 ms), then we may regard it as a false-positive detection or as a false-negative detection. Concerning the metrics for quantifying T-wave and P-wave detection accuracy, the approaches generally apply the same ones used for QRS detection, that is, positive predictivity (P+), sensitivity (Se), and failed detection percentage. The first one is defined as the ratio between the number of true detections and the number resulting from the sum of true detections and false-positive detections. The second one is defined as the ratio between the number of true detections and the number resulting from the sum of true detections and false-negative detections. Finally the failed detection percentage is defined as the ratio between the total number of misdetections and the total number of considered waveforms (for a given ECG record or for a given dataset).

Concerning the evaluation of wave delineation, the literature recommends computing the mean and standard deviation of the time differences between automatic detections and manual expert annotations for T-wave peak, onset and end locations, and for P-wave peak, onset and end locations. Most of the approaches work on a single channel basis, that is, they do not consider multiple detections using several channels for correcting eventual discrepancy and establishing a unique reference. However, depending on the specific database applied for computing simulation tests, two or more leads are available for each ECG record, and the manual annotation process is performed by experts having in sight all available leads. Thus in order to compare—in a reasonable way—manual and automatic detections, a recurring strategy is to choose for each fiducial point (onset, peak or end), and for each specific waveform, the channel with the lowest error. In this regard, the QTDB, which was developed for validating delineation algorithms for ECG characteristic waves, provides 105 ECG records, sampled at 250 Hz and during, each one 15 minutes in duration, and each containing two available leads (Laguna et al., 1997). This database includes records from several other databases pertaining to PhysioNet system, such as MIT–BIH Arrhythmia Database, European ST–T database, ST Change, Supraventricular Arrhythmia, Normal Sinus Rhythm, Sudden Death and Long Term (Goldberger et al., 2000). At QTDB, at least 30 beats per record are manually annotated by ex-

Table 6.1 Performance comparison of T-wave detection rates, T-wave peak, and T-wave end time estimation errors: application to QTDB (NA: not applicable, NR: not reported, TE: time error)

Approach	(Se %)	(P+ %)	T-wave peak: TE (ms)	T-wave end: TE (ms)
Madeiro et al. (2013)	99.32	99.47	1.4 ± 9.0	2.8 ± 15.3
Martínez et al. (2010)	99.20	99.01	5.3 ± 12.9	5.8 ± 22.7
Ghaffari et al. (2009)	99.87	99.80	0.3 ± 4.1	0.8 ± 10.7
Martínez et al. (2004)	99.77	97.79	0.2 ± 13.9	-1.6 ± 18.1
Vila et al. (2000)	92.6	NR	-12 ± 23.4	0.8 ± 30.3
Zhang et al. (2006)	NR	NR	NA	0.31 ± 17.43
Madeiro et al. (2017) (kernel $G_{\sigma_1}[k]$)	99.86	99.97	-5.12 ± 11.46	1.32 ± 14.86
Madeiro et al. (2017) (kernel $G_{\sigma_1,\sigma_2}[k]$)	99.86	99.97	-5.12 ± 11.46	4.49 ± 14.32
Madeiro et al. (2017) (kernel $R_{\sigma_1}[k]$)	99.86	99.97	-5.12 ± 11.46	1.56 ± 15.46
Madeiro et al. (2017) (kernel $R_{\sigma_1}^m[k]$)	99.86	99.97	-5.12 ± 11.46	11.58 ± 17.92

perts, with annotations including peaks, onsets, and ends of T-waves and P-waves. It also includes, for 11 out of its records, an additional set of annotations, which was performed by a second cardiologist expert, totalizing 404 beats with two reference annotations. The CSE multilead measurement database (CSEDB) includes, for each ECG record, 12 standard leads and the Frank leads, totalizing 15 different signals, which should be processed. Therefore 15 different sets of waveform peak and edge detections (results) may be produced for each record by a given delineation algorithm, which is tested over CSEDB. Martínez et al. (2004) establishes, especifically, for CSEDB, a methodology for defining a unique wave edges reference. So after processing each individual lead, they order the 15 single-lead set of detections and select as the reference of edge of a wave the first sample (for wave onset) or the last sample (for wave end) whose k nearest neighbors are within a δ-ms interval. We may find suggested values for k and δ, depending on the analyzed waveform: $k = 3$, and $\delta = 6$ ms for P-wave limits, and $\delta = 12$ ms for T-wave end (Laguna et al., 1994).

Concerning the results computed for a whole dataset, in general the approaches compute the overall average of the errors (m) (time differences between automatic detections and referee annotations) considering all the waveforms from all the records, and the overall average standard deviation by averaging the intrarecording standard deviations.

Thus in order to make a comparison study, we firstly rank a subset of approaches and their corresponding results, including some of the most current reported methodologies, regarding the detection and delineation of T-waves over QT Database, as synthesized within Table 6.1.

It is important to emphasize that the approach developed by Madeiro et al. (2017) adapts the approach proposed by Martínez et al. (2004) concerning the estimation of T-wave peak and P-wave peak locations. Firstly, the approach performs T-wave peak estimation. Then taking this fiducial point as reference (anchor point), the technique applies a given mathematical model for modeling T-wave waveform: Gaussian function ($G_{\sigma_1}[k]$), composition of two Gaussian functions ($G_{\sigma_1,\sigma_2}[k]$), Rayleigh probability density function ($R_{\sigma_1}[k]$), or modified Rayleigh probability density function ($R_{\sigma_1}^m[k]$). Once a fitted kernel is obtained, the methodology allows detecting onset and offset positions, based on strategic sample points derived from the obtained kernel. Therefore the time errors related to T-wave peak detection are the same for each specific kernel, because this step is performed before the modeling task. However the kernel applied for modeling T-wave impacts the accuracy for T-wave delineation (onset and end), and also the delimitation of the search window for detecting and delineating P-wave.

From the data observed within Table 6.1, we can conclude that the detection of T-wave waveform already presents a very high accuracy, that is, the sensitivity and positive predictivity rates do not show great differences and all the algorithms present accuracy rates of around or above 99%. However the T-wave end remains as the most challenging detection problem, especially if we consider the recommendations of the CSE working party (Party, 1985), which are derived from the measurements made by various experts. In this regard, the acceptable standard deviation error value for T-wave end is 30.6 ms (loose criterion) and 15.3 ms (strict criterion).

Now we rank a subset of approaches and their corresponding results, including some of the most current reported methodologies, regarding the detection and delineation of P-waves over QT Database, as synthesized within Table 6.2.

From the data observed within Table 6.2, we observe that the kernel modified Rayleigh probability density function ($R_{\sigma_1}^m[k]$), when applied for modeling both T-wave and P-wave morphologies, provides the highest accuracy for detecting and delineating P-waves. Therefore using $R_{\sigma_1}^m[k]$ for modeling and delineating T-wave allows establishing a search window for investigating P-wave occurrence with the lowest impact degree over the P-wave searching process.

Regarding P-wave detection and delineation, we verify, as for T-wave and QRS complex, that most of the current detectors already present significant high accuracy, providing almost negligible errors. Given that P-wave is commonly a waveform with low amplitudes and easily masked by several types of artifacts, its detection and segmentation require a robust preprocessing for ensuring confidence. It is obvious that for some clinical cases detecting and delineating these waveforms will be specially difficult and challenging. Examples of such cases include when we desire to detect and delineate T-wave morphologies within ECG records presenting atrial fibrillation (the f-waves pollute T-wave edges), and when we desire to detect and delineate P-waves within ECG collected when people are doing physical activity.

Additionally, as we explained in the previous chapter, Jane et al. (1997), Vila et al. (2000), Beraza and Romero (2017) proposed to classify the output of a given algorithm when applied within a given ECG record into one of 4 groups, depending on the obtained parameters mean (μ), or bias, and standard deviation (σ) of the errors: Group 1, acceptable μ and σ; Group 2, acceptable σ but non-acceptable μ; Group 3, acceptable μ but non-acceptable σ; and Group 4, non-acceptable μ and σ. According to Sörnmo and Laguna (2005), an acceptable μ error is considered to be 15 ms and

Table 6.2 Performance comparison of P-wave detection rates, P-wave peak, onset and end time estimation errors: application to QTDB (NA: not applicable, NR: not reported, TE: time error)

Approach	(Se %)	(P+ %)	P-wave onset: TE (ms)	P-wave peak: TE (ms)	P-wave end: TE (ms)
Martínez et al. (2010)	98.65	97.52	2.6 ± 14.5	32 ± 25.7	0.7 ± 14.7
Ghaffari et al. (2009)	99.46	98.83	−1.2 ± 6.3	4.1 ± 10.5	0.7 ± 6.8
Martínez et al. (2004)	99.87	91.03	2.0 ± 14.8	3.6 ± 13.2	1.9 ± 12.8
Di Marco and Chiari (2011)	98.15	91.00	−4.5 ± 13.4	−4.7 ± 9.7	−2.5 ± 13.0
Madeiro et al. (2017) (kernel $G_{\sigma_1}[k]$)	97.07	99.73	−5.73 ± 15.17	−4.04 ± 11.19	−5.63 ± 14.67
Madeiro et al. (2017) (kernel $G_{\sigma_1,\sigma_2}[k]$)	97.72	99.73	−6.95 ± 14.43	−4.0 ± 11.13	−5.26 ± 14.35
Madeiro et al. (2017) (kernel $R_{\sigma_1}[k]$)	96.88	99.73	−15.46 ± 16.84	−4.04 ± 11.19	−6.84 ± 15.38
Madeiro et al. (2017) (kernel $R_{\sigma_1}^m[k]$)	97.76	99.73	−4.23 ± 14.84	−3.95 ± 11.12	2.26 ± 13.14

the acceptable standard deviation depends on each fiducial point. According to Party (1985), the acceptable standard deviation for P-wave onset is 10.2 ms, for P-wave end is 12.7 ms, and for T-wave end is 30.6 ms. In this connection, some delineation approaches perform a record-by-record classification and account for the percentage of records classified within each group. Martínez et al. (2004) provides stratification results with three algorithms for T-wave peak and T-wave end related to records from QTDB. By applying their Wavelet transform delineator, the authors found that 77% of the records from QTDB are classified within Group 1 (well-detected recordings), 8% within Group 2, 5% within Group 3, and 10% within Group 4 (morphology identification error together with poor SNR Beraza and Romero, 2017). In their study, Beraza and Romero (2017) implemented and evaluated 9 algorithms on the QTDB and contained the percentage of records classified within Group 1 for each of the ECG fiducial points: P-wave onset, P-wave peak, P-wave offset, QRS onset, QRS offset, T-wave onset, T-wave peak, and T-wave offset. Based on their study, the approaches proposed by Di Marco and Chiari (2011) and Vázquez-Seisdedos et al. (2011) allow for the highest percentages of records pertaining to Group 1 as it relates to T-wave end detection: 77.78%. For P-wave onset estimation, the approach proposed by Hughes et al. (2004) allows for the highest percentage of records pertaining to Group 1 as it relates to P-wave onset detection: 57.89%.

Finally, given our emphasis for QRS complex detection and segmentation, we highlight some research, which are yet in high demand regarding P-wave and T-wave feature extraction (not fully answered questions). Beyond the classical parameters (amplitude, polarity, duration of P/T-waves, PR interval, QT interval), which characteristics related to the waveform morphology may be useful for recognizing or predict important events? How to recognize subtle changes within waveform, which do not necessarily impact amplitude and duration? Is it possible to condense multi-channel information in a single information signal? Are the most robust methods (according to literature) fast and feasible in real-time applications? Are these methods applicable for smartphone environment and other battery-driven devices?

6.5 CONCLUSIONS

In this chapter, we bore witness to the motivation and the importance for providing accurate automatic detections of T-wave and P-wave. The inherent difficulties are well known, especially considering the low amplitude of the corresponding waveforms, the noise interference, and morphological diversity. We looked at different approaches applied for P- and T-wave detection and segmentation, based on wavelet-transform, nonlinear transformation and mathematical modeling. The Wavelet transform provides a robust band-pass filtering and may be implemented as a computing tool for estimating the location of waveform peaks. We saw that mathematical modeling represents an alternative tool for extracting a wide diversity information from these waveforms, and also learned about the predominant waveforms. Finally, we highlighted research yet in high demands relating to P-wave and T-wave feature extraction, bearing on characteristics related to the waveform morphology, which may be useful for recognizing or predicting important events, and the feasibility of the most current and robust methods in real-time applications.

REFERENCES

Beraza, I., Romero, I., 2017. Comparative study of algorithms for ECG segmentation. Biomedical Signal Processing and Control 34, 166–173.

Di Marco, L.Y., Chiari, L., 2011. A wavelet-based ECG delineation algorithm for 32-bit integer online processing. BioMedical Engineering OnLine 10 (1), 23.

Dilaveris, P.E., Gialafos, J.E., 2002. Future concepts in P wave morphological analyses. Cardiac Electrophysiology Review 6 (3), 221–224.

Ghaffari, A., Homaeinezhad, M., Akraminia, M., Atarod, M., Daevaeiha, M., 2009. A robust wavelet-based multi-lead electrocardiogram delineation algorithm. Medical Engineering & Physics 31 (10), 1219–1227.

Goldberger, A.L., Amaral, L.A., Glass, L., Hausdorff, J.M., Ivanov, P.C., Mark, R.G., Mietus, J.E., Moody, G.B., Peng, C.K., Stanley, H.E., 2000. PhysioBank, PhysioToolkit, and PhysioNet. Circulation 101 (23), e215–e220.

Hall, J.E., 2011. Guyton and Hall Textbook of Medical Physiology. Saunders Elsevier, Philadelphia.

Hughes, N.P., Tarassenko, L., Roberts, S.J., 2004. Markov models for automated ECG interval analysis. In: Advances in Neural Information Processing Systems, pp. 611–618.

Jane, R., Blasi, A., García, J., Laguna, P., 1997. Evaluation of an automatic threshold based detector of waveform limits in Holter ECG with the QT database. In: Computers in Cardiology 1997. IEEE, pp. 295–298.

Laguna, P., Jané, R., Caminal, P., 1994. Automatic detection of wave boundaries in multilead ECG signals: validation with the CSE database. Computers and Biomedical Research 27 (1), 45–60.

Laguna, P., Mark, R.G., Goldberg, A., Moody, G.B., 1997. A database for evaluation of algorithms for measurement of QT and other waveform intervals in the ECG. In: Computers in Cardiology 1997. IEEE, pp. 673–676.

Madeiro, J.P., Cortez, P.C., Marques, J.A., Seisdedos, C.R., Sobrinho, C.R., 2012. An innovative approach of QRS segmentation based on first derivative, Hilbert and Wavelet Transforms. Medical Engineering & Physics 34 (9), 1236–1246.

Madeiro, J.P., Nicolson, W.B., Cortez, P.C., Marques, J.A., Vázquez-Seisdedos, C.R., Elangovan, N., Ng, G.A., Schlindwein, F.S., 2013. New approach for T-wave peak detection and T-wave end location in 12-lead paced ECG signals based on a mathematical model. Medical Engineering & Physics 35 (8), 1105–1115.

Madeiro, J.P.V., dos Santos, E.M.B.E., Cortez, P.C., Felix, J.H.S., Schlindwein, F.S., 2017. Evaluating Gaussian and Rayleigh-based mathematical models for T- and P-waves in ECG. IEEE Latin America Transactions 15 (5), 843–853.

Martínez, A., Alcaraz, R., Rieta, J.J., 2010. Application of the phasor transform for automatic delineation of single-lead ECG fiducial points. Physiological Measurement 31 (11), 1467.

Martínez, A., Alcaraz, R., Rieta, J.J., 2015. Gaussian modeling of the P-wave morphology time course applied to anticipate paroxysmal atrial fibrillation. Computer Methods in Biomechanics and Biomedical Engineering 18 (16), 1775–1784.

Martínez, J.P., Almeida, R., Olmos, S., Rocha, A.P., Laguna, P., 2004. A wavelet-based ECG delineator: evaluation on standard databases. IEEE Transactions on Biomedical Engineering 51 (4), 570–581.

Party, C.W., 1985. Recommendations for measurement standards in quantitative electrocardiography. European Heart Journal 6 (10), 815–825.

Pastore, A.C., Samesima, N., Tobias, N.M.M.O., Pereira Filho, H.G., 2016. Eletrocardiografia Atual: Curso do Serviço de Eletrocardiografia do Incor. Atheneu, Rio de Janeiro.

Platonov, P.G., 2012. P-wave morphology: underlying mechanisms and clinical implications. Annals of Noninvasive Electrocardiology 17 (3), 161–169.

Sörnmo, L., Laguna, P., 2005. Bioelectrical Signal Processing in Cardiac and Neurological Applications, vol. 8. Academic Press.

Surawicz, B., Knilans, T., 2008. Chou's Electrocardiography in Clinical Practice E-Book: Adult and Pediatric. Elsevier Health Sciences.

Vázquez-Seisdedos, C.R., Neto, J.E., Reyes, E.J.M., Klautau, A., de Oliveira, R.C.L., 2011. New approach for T-wave end detection on electrocardiogram: performance in noisy conditions. BioMedical Engineering OnLine 10 (1), 77.

Vila, J.A., Gang, Y., Presedo, J.M.R., Fernández-Delgado, M., Barro, S., Malik, M., 2000. A new approach for TU complex characterization. IEEE Transactions on Biomedical Engineering 47 (6), 764–772.

Zhang, Q., Manriquez, A.I., Médigue, C., Papelier, Y., Sorine, M., 2006. An algorithm for robust and efficient location of T-wave ends in electrocardiograms. IEEE Transactions on Biomedical Engineering 53 (12), 2544–2552.

The Issue of Automatic Classification of Heartbeats

Priscila Rocha Ferreira Rodrigues*, José Maria da Silva Monteiro Filho*,
João Paulo do Vale Madeiro†

*Department of Computing Science, Federal University of Ceara, Fortaleza, Ceará, Brazil
†Institute for Engineering and Sustainable Development (IEDS), University for the International
Integration of the Afro-Brazilian Lusophony – UNILAB, Redenção, Ceará, Brazil

7.1 ARRHYTHMIA CLASSIFICATION PROCESS

The incidence of cardiovascular disease is one of the leading causes of death worldwide, making it imperative to search for new methods of early diagnosis and more effective treatments, as well as the prevention of diseases related to the heart. According to Benjamin et al. (2017), the increase in the number of cardiology patients in the world is directly related to changes in modern lifestyle. Many of these victims have considered unhealthy behaviors, such as smoking, overeating of food, and inappropriate physical activity.

Through the aid of preventive examinations, the great majority of heart diseases can be diagnosed at an early stage and treated in advance. The most commonly used diagnostic test for heart disease is the electrocardiogram (ECG). As a result, numerous researches have sought to extract patterns from ECG signals databases so that such diseases, such as arrhythmias, can be identified effectively.

Arrhythmia is the generic name for several disorders that change the frequency and/or rhythm of heartbeats. They can occur for several reasons. Arrhythmias can lead to death and therefore constitute a medical emergency. Most of them are, however, harmless. The sinus node in the right atrium is a group of cells that regulate these beats by electrical impulses that stimulate the contraction of the heart muscle or myocardium. When these electrical impulses are emitted irregularly or conducted poorly, cardiac arrhythmia may occur. This can be characterized by excessively rapid rhythms (tachycardia), slow (bradycardia) or only irregular rhythms (Hall, 2017). To be considered normal, the heart rate for a resting individual should be around 60 to 100 beats per minute. Arrhythmias are characterized by changes in wave morphology, thus forming a pattern. From this pattern, it is possible to identify and classify the type of arrhythmia. However, it is very difficult and tiring for medical professionals to analyze long ECG records in a short period of time, in addition, the human eye is

Developments and Applications for ECG Signal Processing. https://doi.org/10.1016/B978-0-12-814035-2.00013-X

inappropriate to detect the morphological variation of the ECG signal, thus imposing the need for the use of computational techniques for automatic classification.

Although various types of cardiac arrhythmias exist, the American National Standards Institute (ANSI) and the Association for the Advancement of Medical Instrumentation (AAMI) (1998) has developed a norm (ANSI/AAMI EC57: 1998-R 2008) published in 1998 and updated in 2008 (ANSI/AAMI, 2008) regulating which arrhythmias should be detected by the algorithms intended for this purpose. As it is described in Table 7.1, the norm recommends the classification of the beats in 15 classes, divided into 5 groups (super-classes): normal beat, left bundle branch block (LBBB), right bundle branch block (RBBB), and atrial escape and nodal junction escape beats belong to group N category; group S contains atrial premature (AP), aberrated premature (aAP), nodal junction premature (NP), and supraventricular premature (SP) beats; group V contains premature ventricular contraction (PVC), and ventricular escape beats (see Fig. 7.1), group F contains only fusion of ventricular and normal (fVN) beats, and group Q, which is known as unknown beat, contains paced beat (P), fusion of paced and normal (fPN) beats, and unclassified beats. A sample ECG graph of each of the five classes of arrhythmias is shown in Fig. 7.1.

The process for automating the arrhythmia classification using ECG signals can be divided into four sequential steps (see Fig. 7.2): pre-processing, segmentation, feature extraction, and classification. At the end of each step an action will have been taken with the ultimate goal of determining the type of heartbeat (Luz et al., 2016).

Preprocessing is the step, where the ECG signal passes through a filtering to eliminate noise that hinders its analysis. These noises can originate from the electrical network, through incorrect placement of the electrodes, or even by the movement of the patient. Among the many proposals to reduce noise in ECG signals, the most simple and widely used is the implementation of recursive digital filter response to the finite impulse (FIR) (Lynn, 1971), made computationally viable with the advancement of microcontrollers and microprocessors. Techniques for preprocessing the ECG signal are extensively explored, however, choosing which method to use is directly linked to the final research goal. Methods focused on the segmentation of the ECG signal (detection of the QRS complex) tend to require different preprocessing when compared to the methods for the automatic classification of arrhythmias. Chapter 3 details this stage of preprocessing.

In the segmentation stage (see Chapter 5), the QRS complex is delimited, which reflects the electrical activity of the heart during the ventricular contraction. Once the segmentation is performed, several physiological information can be obtained, such as heart rate, pulse width, amplitude, among others. From this information, techniques for extracting characteristics, which will serve as a basis for automatic classification, may be applied. The importance of this step in the classification of arrhythmias is emphasized, since errors in the detection of the QRS complex will be propagated, and will strongly impact the classification.

The characteristics to be used in the classification phase of ECG signal arrhythmias are crucial to the success of the model. Therefore the feature extraction step is an extremely relevant point because it will directly interfere with the performance

Table 7.1 ECG class description using AAMI standard

GROUP	CLASS				
Normal beat (N)	Normal beat (N)	Left bundle branch block beat (L)	Right bundle branch block beat (R)	Atrial escape beat (e)	Nodal (junctional) escape beat (j)
Supraventricular Ectopic Beat (SVEB)	Atrial premature beat (A)	Aberrated atrial premature beat (a)	Nodal (junctional) premature beat (J)	Supraventricular premature beat (S)	
Ventricular Ectopic Beat (VEB)	Premature ventricular contraction (V)	Ventricular escape beat (E)			
Fusion beat (F)	Fusion of ventricular and normal beat (F)				
Unknown beat (Q)	Paced beat (/)	Fusion of paced and normal beat (f)	Unclassified beat (Q)		

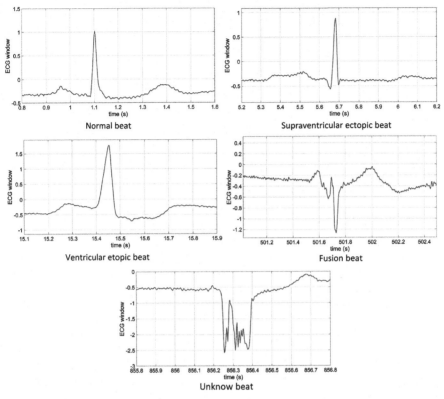

FIGURE 7.1

Five classes of ECG beat.

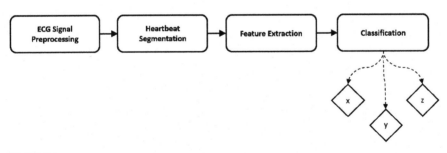

FIGURE 7.2

Stages of the arrhythmia classification process.

and accuracy of the entire process. All information extracted from a heartbeat, which can be used to determine its type is considered a characteristic. The features can be extracted in different ways, directly from the wave morphology of the ECG, both in time domain and in the frequency or heart rate (also known as RR interval).

The last step, classification, is where arrhythmias are identified indeed, according to the classes arranged in the ANSI/AAMI standard. Once the feature set has been defined, it is possible to construct models for automatic arrhythmia classification using machine learning algorithms.

The last two steps of the classification process, feature extraction and classification, will be discussed in the second and third sections of this chapter, respectively.

7.2 FEATURE EXTRACTION

The feature extraction step is the key to the success in the heartbeat pattern recognition using the ECG signal. A feature can be defined as any information extracted from the heartbeat used to discriminate it. The features can be extracted in many ways from the ECG signal's morphology, in different domains: the cardiac rhythm, the time domain, the frequency domain, and the time-frequency domain. Most popular methods proposed in literature are discussed next.

7.2.1 CARDIAC RHYTHM

Initially, the most simple feature found in the literature are known as the RR interval. It is calculated from the cardiac rhythm (or heartbeat interval). The RR interval can be defined as the time between the R-peak of a heartbeat with respect to another heartbeat, which could be its predecessor or successor. Variations in the width of the RR interval are, in general, correlated with the alterations in the morphology of the curve, frequently provoked by arrhythmias (Luz et al., 2016). For this reason, the RR interval is a highly critical factor in the discrimination of heartbeat types. Thus some authors have based their methods only on using the RR interval (Exarchos et al., 2007; Kumar and Kumaraswamy, 2013). Variations of RR interval are used to reduce noise interference and are very common, example, the average of the RR interval for a certain period (Ye et al., 2010).

Lin and Yang (2014) showed that the use of normalized RR-intervals improves significantly the classifiers accuracy. In the same vein, Doquire et al. (2011) proved the efficiency of normalized RR-intervals by means of feature selection algorithms.

Other features extracted from the heartbeat intervals are also found in literature, such as the ECG-intervals (or ECG segments): PR Interval, QRS Interval, QT Interval, among others. The QRS Interval, the duration of the QRS complex, is the most utilized, since some types of arrhythmias provoke variations in the QRS Interval. Thus the QRS Interval is a good discriminating feature (de Chazal et al., 2004; Korürek and Doğan, 2010).

At their book, Gari et al. (2006) report standard values and tolerance bands for P-wave duration, PQ/PR interval, QRS width, QT interval, amplitude of P-waves, amplitude of QRS complex, ST-level and T-wave amplitude, considering a healthy human being with no cardiac abnormalities.

7.2.2 TIME DOMAIN

The Pan–Tompkins algorithm (Pan and Tompkins, 1985) is most widely used and highly acknowledged technique for ECG feature extraction. It is a simple yet powerful algorithm for real-time QRS detection. The Pan–Tompkins algorithm is based on the slope, amplitude, and width of the QRS complex. It uses a special digital band-pass filter in order to reduce the noises and false detection caused by faulty signals that contaminate the ECG signal. The Pan–Tompkins algorithm is simple since it automatically adjusts the threshold value for next peak detection from the previous smallest peak. In a simplified way, this algorithm goes through the following steps (Agrawal and Gawali, 2017):

- Band-pass filtering;
- Applying a derivative operator;
- Square and integrate signal;
- Fiducial mark determination by thresholding;
- QRS wave detection;
- Average RR interval and rate limits adjustment.

Valluraiah and Biswal (2015) proposed a method, where the ECG signal is preprocessed and is subjected to Hilbert transform along with a window to enhance the presence of QRS complexes, to detect R-peaks in the ECG waveform by setting a threshold. This algorithm utilizes the first differentiation of the ECG signal. This approach has a few advantages over the other methods of detecting QRS. First, it can minimize the unwanted effect of the large T- and P-peaks by creating envelope around the R-peak. Besides using an odd function the method avoids the signal crossing the X axis whenever there is any kind of disturbance.

The linear prediction method is used in audio signal and speech processing in order to reproduce the original signal accurately. Sarma et al. (2013) used Linear Prediction Coefficients (LPC) for extracting ECG features.

A histogram is a representation of the probability distribution of a continuous variable (quantitative variable) and was first introduced by Karl Pearson. Halder et al. (2016) proposed a histogram-based method for detection and identification of R-wave, P-wave and T-wave from noise removal ECG signal. In this work, histograms are used as an estimator of the waves of a ECG signal. For this, the whole ECG signal is divided into few small windows of predefined width. The histograms are generated by measuring the variations of the orientations among these sample values in some quantized directions. Next, few zones are depicted as QRS zones, having the area more than a predefined threshold. The local maxima of these zones are considered as the R-peak. Based on the same strategy, P- and T-waves can be detected. This method has the advantage of being used directly for online analysis without using any complex mathematical models.

The simplest way to extract features in the time domain from the ECG signal curve is using its own sampled points as features (Özbay and Tezel, 2010; Luz et al., 2013). However, using samples from the ECG curve as features is not an effective technique for two reasons. First, the dimension of the feature vector produced is high, since it depends on the amount of samples used to represent the heartbeat. Second, it suffers with several drawbacks regarding scale and displacement related to the central point (the R-peak). With the purpose of decreasing the feature vector size and avoid the aforementioned problems, some approaches have been proposed, such as interpolation, principal component analysis (PCA), independent component analysis (ICA), and generalized discriminant analysis (GDA).

de Chazal et al. (2004) used the interpolation of the ECG signal, such that the final time representation is composed of 18 and 19 samples obtained from 250 samples. Thus this process is applied in the two leads available in the dataset used. Next, a feature vector composed of the 52 best features reported is used.

Chawla (2008) presented a comparative analysis between the use of PCA and ICA to reduce the noise and artifacts of the ECG signal and showed that PCA is a better technique to reduce noise, whereas ICA is better to extract features. The ECG is a mix of several action potentials. The ICA technique allows statistically separating individual sources from a mixed signal. The PCA technique separates the sources according to the energy contribution to the signal. Thus the combination of these two techniques, that is, PCA for noise reduction and ICA for feature extraction, can offer greater advantages when compared to using only one of them.

Kanaan et al. (2011) investigated another technique based on PCA, the Kernel Principal Component Analysis (KPCA). The authors performed a comparison between PCA and KPCA and they concluded that KPCA is superior to the PCA technique for classifying heartbeats from the ECG signal.

Asl et al. (2008) used Generalized Discriminant Analysis (GDA) in order to reduce the dimensions of the features of the heartbeat. Initially, 15 different features were extracted from the input signal using linear and nonlinear methods. Then these features were reduced to only five features by the GDA technique. It is important to highlight that this not only reduces the number of input features, but also selects the most discriminating features.

One of the most important methods of detection of the R-wave peak, and others, is the usage of adaptive threshold. Chen et al. (2015) proposed an R-peak adaptive threshold extraction algorithm. The main idea behind this algorithm is that the value of threshold changes with time, and it gets updated time-to-time in correspondence to the changes in the QRS morphology, and the levels of noise present in the ECG signal. This method is widely used to detect the peaks and the baseline drift and power interference, since it is much simpler and more effective. Besides, it can also do the real-time ECG processing. The adaptive threshold algorithm is simple, accurate, real-time performing and stable. It can also effectively solve the baseline drift problems.

7.2.3 FREQUENCY DOMAIN

The frequency domain methods work as alternative tools to allow identifying subtle variations in amplitude and morphology of different waveforms, even in real scenarios, with noise interference, baseline wandering, and artifacts. Actually, frequency based methods help to explore hidden information (Martis et al., 2014). In this sense, Fourier transform and power spectral density (PSD) are two basic techniques, which are recurrently applied for feature extraction. According to Khazaee and Ebrahimzadeh (2010), power spectral estimation is still one of the most widely-used methods of signal analysis, being related to the correlation function through the Fourier transform. Basically, spectral estimation is applied to describe the distribution of the power embedded in a signal over frequency. The more correlated or predictable a signal, the more concentrated its power spectrum. As an inverse analogy, the more unpredictable a signal, the more widespread its power spectrum.

According to Vaseghi (2008), Khazaee and Ebrahimzadeh (2010), the methods for spectral estimation may be classified as: nonparametric estimation, model-based estimation, and high-resolution estimation based on subspace eigen-analysis.

The Fourier transform $X(f)$ of a discrete-time signal $x[n]$, considering a signal window with N samples, is given as (Proakis, 2001)

$$X(f) = \sum_{n=0}^{N} x[n]e^{-j2\pi fn}. \qquad (7.1)$$

The power spectral density may be estimated from the Fourier transform using the expression

$$P_x(f) = \frac{1}{N}|X(f)|^2. \qquad (7.2)$$

We also may estimate the power spectral density as the Fourier transform of an autocorrelation function computed within a signal $x(t)$. Thus the power spectral density for a stationary stochastic process $x(t)$ is defined as

$$P_x(j\omega) = \int_{-\infty}^{\infty} R_X(\tau)e^{-j\omega\tau}d\tau, \qquad (7.3)$$

where $R_X(\tau)$ is the autocorrelation function applied within a signal $x(t)$.

If we define $X_T(t)$ as a segment or subsequence of a stochastic process or signal $X(t)$, during T, we may define the periodogram for each particular realization of $X_T(t)$ as

$$Periodogram[X_T(t)] = \frac{1}{T}|F[X_T(t)]|^2, \qquad (7.4)$$

which coincides with Eq. (7.2).

The expected value of the periodogram as $T \rightarrow \infty$ provides a nonparametric estimation of PSD. By using the Welch method, we can reduce the variance of that estimator through a simple process: segmentation of the signal $X_T(t)$ into K adjacent and overlapping subsequences, each one with N samples, $x_N^{(k)}, 0 \le k \le K - 1$; application of a window function over each subsequence for providing different weights for each sample within a subsequence; and, finally, the computing of a modified periodogram for each subsequence, given as

$$\hat{S}_N^k(j\omega) = \frac{1}{N}|\sum_{t=0}^{N-1}\omega(t)x_N^{(k)}(t)e^{-j\omega.t}|^2. \tag{7.5}$$

The window function $\omega(t)$ has the propriety

$$\frac{1}{N}\sum_{t=0}^{N-1}\omega^2(t) = 1. \tag{7.6}$$

Then the estimation of the power spectral density for a signal $X_T(t)$ is computed as the average of the periodograms $\hat{S}_N^k(j\omega)$ of the respective subsequences.

If we represent the signal or stochastic process $X(t)$ as the output generated by a linear filter $H(z)$, whose input is a white noise process $\varepsilon \sim N(0, \sigma_{\varepsilon^2})$, we obtain an ARMA process (autoregressive and moving average), given as (Shynk, 2012)

$$x(t)+a_1x(t-1)+...+a_px(t-p) = \varepsilon(t)+b_1\varepsilon(t-1)+...+b_q\varepsilon(t-q). \tag{7.7}$$

If we define a delay operator $z^{-d} = e^{jd\omega}$, then Eq. (7.7) may be written as $A(z)X(z) = B(z)\varepsilon(z)$ (z-transform domain). It follows that $H(z) = B(z)/A(z)$. It is possible to demonstrate that the power spectral density of a signal or stochastic process $x(t)$ properly modeled by an ARMA process is given as

$$S_X(j\omega) = \sigma_\varepsilon^2|H(e^{j\omega})|^2. \tag{7.8}$$

If we replace $H(e^{j\omega})$ by $1/A(e^{j\omega})$, then we find an expression for estimating the power spectral density of a signal modeled by an AR (autoregressive) process:

$$S_X^{AR}(j\omega) = \frac{\sigma_\varepsilon^2}{|A(e^{j\omega})|^2}. \tag{7.9}$$

Martis et al. (2014) reviewed analysis methods in electrocardiogram characterization, including the tools for frequency transform domain. In their survey, the authors highlighted the approach developed by Khazaee and Ebrahimzadeh (2010). The authors chose to use nonparametric power spectral density estimation methods in an ECG beat classification problem. The proposed technique, named power spectral-based hybrid genetic algorithm-support vector machines (SVMGA) allows classifying five types of beats: normal beats and four arrhythmia classes (LBBB, RBBB, APC, and PVC). For classifying and recognition task, the authors employed support vector machine (SVM).

7.2.4 TIME-FREQUENCY DOMAIN

Discrete wavelet transform (DWT) is sort off called wavelet filter banks, as it uses two filters, a low-pass Filter (LPF), and a high-pass filter (HPF) to decompose the signal into different frequency scales. This technique divides the whole wave, that is, the mother wavelet into different wavelet transform bases. The DWT, however a powerful tool for ECG feature extraction, is still limited by the choice of wavelets for use in feature extraction process. Mukherjee and Ghosh (2012) presented a comparative study on the choice of some of the conventionally used wavelets and their corresponding accuracies for ECG feature extraction by R-peak detection. In this context, the B-Spline wavelet transform of ECG signal is found to be the most efficient means of ECG signal processing. Lim et al. (2015) proposed an adaptive signal extraction method that uses Discrete Wavelet Transformation, coupled with adaptive parameters to address variated heartwave signal due to varying heartrates. In this paper, the features of the heart wave signal comprises of the P-wave, QRS-complex, T-wave, onset and offset of P-wave and T-wave. Besides using statistically deduced parameters, the PR and QT interval parameters were incorporated, where the signal extraction method can be made adaptive to varying heartrate that resulted in a very reliable signal extraction method. da Silva (2015) presented the application of wavelet transform for characterization of QRS complex in ECG signals and removal noise. Digital signal processing techniques are applied as the Wavelet transform to enable the analysis of that signal.

Independent component analysis (ICA) is a method used in signal processing in order to separate a multivariate signal into its subcomponents, where these signals are mutually independent. Thus ICA can be used for the removal of such noises and artifacts. Phegade and Mukherji (2013) applied different ICA schemes, such as JADE algorithm and Fast ICA for ECG denoising. Besides, the authors also made an attempt to apply constrained ICA (cICA) for ECG signal denoising, which was used for fetal ECG extraction.

Auto-regression (AR) (or Autoregressive) method is a two-lead ECG model. It extracts the needed features from the acquired ECG signal and classify them according to disease. The auto-regression coefficients computed from the ECG signal are classified using a generalized linear model and a multilayer feed forward neural network. The AR method has a lead over the other methods based on its simplicity, and also how well-suited it is for real-time applications. Besides, AR has greater accuracy (Vaneghi et al., 2012).

Rodríguez et al. (2015) presented an approach for QRS complex detection and extraction of ECG signals based on adaptive threshold and principal component analysis. Initially, the ECG signal is filtered by a band pass filter, and then it is differentiated. Next, the Hilbert transform and the adaptive threshold technique are applied for QRS detection. Finally, the Principal Component Analysis is used to extract features from the ECG signal.

Ye et al. (2010) combined several ECG feature extraction techniques found in the literature to generate their feature vector. In this paper, the features are extracted from 300 samples surrounding the R-peak; 100 before and 200 after the R-peak. Next, the wavelet transform is applied to the sampled ECG signal and 118 coefficients are extracted. Besides, other eighteen coefficients are extracted using the ICA. The authors call this set of feature "morphological". With the purpose of reducing the dimension of the obtained morphological feature vector to 26, the authors employed the principal components analysis (PCA) technique.

Karpagachelvi et al. (2010) discussed various techniques and transformations proposed in literature for extracting feature from an ECG signal. Besides, this paper also provides a comparative study of different methods proposed by researchers in extracting features from ECG signal.

Vaneghi et al. (2012) discussed six most frequent methods used to extract different features in ECG signals: Auto-regression (AR), Wavelet Transform (WT), Eigenvector, Fast Fourier Transform (FFT), Linear Prediction (LP), and Independent Component Analysis (ICA). This study revealed that Eigenvector method gives better performance in frequency domain for the ECG feature extraction.

Agrawal and Gawali (2017) performed a comparative analysis among eight feature extraction methods based on various performance parameters, such as Sensitivity, Predictivity, and Accuracy. The investigated methods were Pan–Tompkins, Hilbert transform, Histogram approach, Wavelet transform, Auto-regression (AR), Independent Component Analysis (ICA), Linear prediction (LP), and Adaptive threshold. This paper provided insights for using different methods for ECG feature extraction.

Vincent and Sreekumar (2017) studied the different features that can be extracted from ECG signal mainly based on its waveform components. This paper presented a table, including the main researches done in this field, the methods proposed for ECG feature extraction, the used ECG features, and various cardiac disease classes.

7.3 FEATURE SELECTION

It is important to highlight that, even if some works regard feature extraction and feature selection as two interchangeable terms, these two steps are in fact different. While feature extraction is defined as the process, which involves the description of a heartbeat, feature selection consists in choosing a subset with the most representative features, with the objective to improve the pattern recognition task. Feature selection techniques can bring many benefits to the pattern recognition methods, such as the increase of the generalization power of the learning algorithms and the reduction of the computational cost, since they use a smaller number of features to construct the models (Mar et al., 2011).

In this context, many authors have used techniques that reduce the feature space (using PCA, for example), but few have investigated techniques for feature selection (Llamedo and Martinez, 2011). Llamedo and Martinez (2011) employed, for the first time in literature, an algorithm for feature selection by using floating sequential search. The proposed method interchanges algorithms executing forward and backward searches to obtain a set with the most robust features and avoid local optima in the feature space. This approach achieved better results than the state-of-art methods using only eight selected features.

Mar et al. (2011) also performed feature selection, by using the floating sequential search. In this study, the authors analyzed a set of different possibilities for feature selection, searching for a trade-off between the number of features and accuracy. In addition to the linear discriminant (LD) classifier, used in their previous works (Mar et al., 2011), the authors employed a multilayer perceptron. However, neither of these results were better than those obtained in their previous work (Llamedo and Martinez, 2011) in terms of accuracy.

Doquire et al. (2011) have compared a wrapper feature selection technique against a filter feature selection technique. The wrapper feature selection was used with the weighted LD model using a forward–backward search strategy. The filter technique employed was the mutual information, together with ranking approach and weighted SVM (Support Vector Machines). The results showed that higher figures are obtained when a very small number of features are selected. The authors highlighted that the most important features appear to be R–R intervals. Besides, they claimed that the mutual information criterion is a powerful tool for feature selection in the context of arrhythmia classification.

Zhang et al. (2014) argue that many ECG features are associated with mathematical interpretation and do not have a clear meaning to physicians. Thus the authors proposed a heartbeat class-specific feature selection method to allow the study of feature contribution for each arrhythmia/heartbeat class. This investigation can bring about important insights by allowing better understanding of correlation among heart diseases and features extracted from ECG.

Some other techniques for attribute selection have been proposed in the literature, such as Genetic Algorithms (GA) (Oh et al., 2004) and particle swarm optimization (PSO) (Korürek and Doğan, 2010; Wang et al., 2007; Lin et al., 2008). They can also provide promising results and should be better investigated in future works.

7.4 LEARNING ALGORITHMS

Models for classification of arrhythmias using learning algorithms and data mining techniques can be constructed from a set of previously defined characteristics (Bishop, 2006). Over the years, different authors have applied the most varied types of artificial intelligence algorithms in this context. Such algorithms are being evaluated and compared, with the objective of achieving increasingly optimum results of classification and analysis of the heart rate. The authors employed machine learning methods, statistical, and numerical techniques, even the hybrid approach of these methods. The accuracy of each classification method depends on the situations and the processes that they carried out.

In the following subsections, the most popular learning algorithms in the literature for classification of arrhythmias will be presented and their characteristics discussed in more detail.

7.4.1 **SUPPORT VECTOR MACHINES (SVM)**

SVMs are methods that construct classifiers from the creation of hyperplanes in an n-dimensional space, that is, drawing "lines" in an n-dimensional space that are able to separate examples of different classes (Schölkopf and Smola, 2002). These hyperplanes are created with the aid of points located at the edges of the sets, called support vectors.

When confronted with nonlinear problems, SVMs create a mapping between a set of input values (examples) and a feature space, in which these initial nonlinear boundary classes are made linearly separable by a transformation (or mapping) of the space of features. This mapping is done by a set of mathematical functions called kernels. After performing this mapping, SVMs use an iterative training algorithm to minimize an error function.

For arrhythmia classification using ECG signals, SVM is one of the most popular classifiers found in the literature. However, as few papers reviewed took into account the recommendations of the AAMI standard or were careful not to mix beats of the same patient in the training sets and the test, the vast majority of published studies present unrealistic results from a clinical point of view, and can not be used as a source of comparison. But in Park et al. (2008), the authors used the SVM and validated the method according to the AAMI protocol and the scheme of division proposed by de Chazal et al. (2004), and also used SVM in a pseudo-hierarchical configuration to solve the MIT–BIH Database imbalance.

In the literature there are several other approaches with SVM variations, such as combined genetic algorithms for SVM, and with fuzzy constraints (Nasiri et al., 2009), arrhythmia classification using spectral correlation and support vector machines (Khalaf et al., 2015), low Wavelet resource combination dimension and SVM (Qin et al., 2017), and so on.

7.4.2 **ARTIFICIAL NEURAL NETWORKS (ANN)**

ANNs are computational techniques that present a model based on the neural structure of intelligent organisms that acquire knowledge through experience (Krose and van der Smagt, 1996). The functioning of the biological neuron was interpreted by the physiologist (McCulloch and Pitts, 1943) as being a circuit of binary inputs, combined by a sum weighted (with weights) producing an effective input. The neuron consists of:

- Inputs (dendrites) – where signals are applied;
- Weights (synapses) – where knowledge is retained;
- Sum function – sum of the ratio of input signals and weights synaptic;
- Activation function – function that, depending on the value of the sum, will or will not activate the output depending on the threshold (sigmoidal, ladder, Gaussian, etc.);
- Output (axon) – interface output.

They have good applications, where the rules of solving the problem are unknown or difficult to formalize, and when it is necessary to quickly solve the problem. ANNs are self-adaptive, nonlinear, fast, and accurate. They are also robust to noise and easily scalable. With a network of artificial neurons, it is possible to perform different tasks, such as classification of patterns, predictions, and approximation of functions.

According to Jambukia et al. (2015), among the advantages of using neural networks are:

- They provide nonlinear mapping between inputs and outputs using a sigmoid activation function to solve nonlinear problems, such as ECG Signal classification;
- They are able to achieve better results than statistical or deterministic approaches. Statistical methods work well for linear problems, but fail to produce good results for nonlinear problems developed on the assumption of given time series;
- ANNs can adaptively model the lower frequencies of the ECG that are inherently nonlinear;
- They are capable of removing nonlinear and variable noise characteristics of the signals.

Among the disadvantages of the application of neural networks are:

- ANN training algorithm is not able to guarantee that a global minimum is reached;
- It is possible that ANN does not provide the optimal solution for the entire ECG 12-lead classification process.

Many researchers have used different types of neural networks to classify ECG signals. The most commonly used ANN architectures for arrhythmia classification are Multi-layered Perceptrons (MLP) and Probabilistic Neural Networks (PNN). Yu and Chen (2007) argue that models built with PNN are more computationally efficient than the traditional MLP. However, a hybrid neuro-fuzzy network was proposed in order to increase generalization and reduce the training time of the MLPs (Özbay et al., 2006). The PNN needs little or no training, except spread optimization. The experience of the PNN is limited to very small sets of data. Banupriya and Karpagavalli (2014) shows that the performance of the NNP decreased slightly with the reduction of training data size.

In Khorrami and Moavenian (2010), the same methodology was used to compare results obtained between SVM and MLP–ANN. The MLP performed better in terms of Accuracy, Sensitivity, Positive Predictivity and False Positive Rate, only losing efficiency over time for both training and testing. Mar et al. (2011), compared MLP with Linear Discriminates and also concluded that MLPs performed significantly higher.

More recently, a 9-layer deep convolutional neural network (CNN) has been developed that can automatically identify 5 different categories of heartbeats on ECG signals. The experiments showed that CNN showed excellent accuracy rates even in noisy conditions (Acharya et al., 2017). In Gutiérrez-Gnecchi et al. (2017), a probabilistic neural network is used in an arrhythmia classification method implemented in a Digital Signal Processing (DSP) platform for on-line real-time outpatient operation. An aspect that has not yet been explored is the combination of classifiers, which according to Osowski et al. (2008), can reduce the incidence of false negatives and global errors.

7.4.3 LINEAR DISCRIMINANTS (LD)

Linear Discriminants is a statistical method of dimensionality reduction that provides the highest possible discrimination among various classes, used in machine learning to find the linear combination of features, which can separate two or more classes of objects with best performance. It has been widely used in many applications, such as pattern recognition, image retrieval, speech recognition, among others. The method is based on discriminant functions that are estimated based on a set of data called training set. These discriminant functions are linear with respect to the characteristic vector, and usually have the form

$$f(t) = \mathbf{w}^t x + b_0, \tag{7.10}$$

where w represents the weight vector, x the characteristic vector, and b_0 a threshold.

The criteria adopted for the calculation of the vector of weights may change according to the model adopted. In de Chazal et al. (2004), for example, the parameters were determined by maximum-likelihood calculated from the training data. In terms of standardization, LDs are the classifiers used that most meet the AAMI recommendations, being often chosen by the authors for their simplicity and the emphasis that is wanted on the proposed characteristics (Luz et al., 2016).

One of the most recent classifiers proposed with LD was able to classify and accurately differentiate normal and abnormal heartbeats through a simple, efficient, and rapid method that did not require the use of complex mathematics (Yeh et al., 2009). LDs can easily overcome problems generated by the imbalance of the training set (a difficulty presented by approaches based on SVM). Another advantage is that the LD classifier, when compared to SVM and MLP, requires less training time, because it simply calculates training data statistics, and then the classification model is defined; so it is not iterative.

7.4.4 OTHER TECHNIQUES

In addition to the more popular techniques presented in the previous subsections, many other approaches have been applied to classify arrhythmias. Classifiers using decision tree are relevant because they allow an interpretation of the decisions taken by the model, giving the medical professional greater confidence and analysis to analyze the final results. Methods that apply decision trees use few characteristics, this type of approach being inefficient for continuous characteristics and vectors of high-dimensional characteristics (Exarchos et al., 2007). Hyperbox classifiers, in addition to providing high-level interpretation of classification rules, are also efficient for feature vectors of larger dimensions (Bortolan et al., 2007).

Researches based on k Nearest Neighbors (kNN) are not commonly used in the context of classification of arrhythmias due to the high computational cost presented. Mishra and Raghav (2010) and Lanata et al. (2011) are examples of research using kNN. However, none makes use of the AAMI recommendations, preventing a fair comparison. Korürek and Nizam (2008) presents an efficient arrhythmia clustering algorithm, based not only on the general knowledge of signal detection, but also on the specific features of the ECG signal. In addition, clustering techniques are widely used in conjunction with neural networks, improving generalization capacity and shortening learning time (Lagerholm et al., 2000; Özbay et al., 2006).

Reservoir Computing (RC) models use a logistic regression learning model that has proven to be suitable for hardware implementations because of its low cost, which enables it to be used in applications that classify beats in real time (Escalona-Morán et al., 2015). Since recently, deep learning techniques are being employed to classify ECG signals and the results obtained are promising, showing that the approach provides significant precision improvements, with less interaction and faster online retraining, when compared to state-of-the-art methods (Al Rahhal et al., 2016).

7.5 DATABASES FOR ECG ANALYSIS

Data collections composed of ECG patient records, duly noted, are configured as a source of essential resources for the development and evaluation of automatic classifiers arrhythmias, as well as for the advancement of research in this area.

Earlier researchers in the field typically built their own ECG databases and used them to develop and evaluate their arrhythmia detection programs (Gath and Inbar, 2013). However, the diversity of wave morphologies, heart rhythms, and presence of noise presented constant challenges to algorithm designers, which made evident the need for standardization of these databases, so that classifiers of increasingly sophisticated arrhythmias could be evaluated objectively.

In this context, some educational institutions, allied to hospitals and partnerships with medical professionals, began to develop large structured databases with the purpose of stimulating new research and investigations in the study of cardiovascular diseases, and other health areas. Today, there are a number of widely available databases, including cardiopulmonary, neural, and biomedical signals from both healthy individuals and patients with a variety of health problems, including arrhythmias, heart failure, sleep apnea, and neurological disorders, among others.

The AAMI standard recommends the use of five databases, which are standardized according to the protocol defined by the standard, so that the results of the studies can be properly reported and considered reproducible or comparable (see Table 7.2). The standard also specifies issues, such as the annotations that should be made in the databases and the measures that should be used to evaluate the methods of classification of arrhythmias, which are Sensitivity (Se), Positive Predictivity (+P), False

Table 7.2 Features of standard ECG databases

DATABASE	ACRONYM	RECORDS	RECORD DURATION	AVAILABLE SINCE
The Massachusetts Institute of Technology and Beth Israel Hospital Arrhythmia Database	MI–BIH	48 records	30 min	1980
The American Heart Association Database for Evaluation of Ventricular Arrhythmia Detectors	AHA	80 records	35 min	1982
The European Society of Cardiology ST-T Database	ESC ST–T	90 records	2 h	1991
The Creighton University Sustained Ventricular Arrhythmia Database	CU	35 records	8 min	1986
The Noise Stress Test Database	NST	12 records	30 min	1984

Positive Rate (FPR) and Global Accuracy (Acc). In the literature, Sensitivity and Positive Predictivity are also called Recall and Precision, respectively.

Each of these databases presented in Table 7.2 represents a very substantial effort by numerous researchers, in particular the AHA, MIT, and ESC ST-T Databases that required more than five years of efforts by large teams of researchers and clinicians from many institutions. However, it is important to recognize that even if these banks allow standardized, quantitative, automated, and fully reproducible assessments of analyzer performance, these data do not fully represent the variety of "real-world" ECGs observed in clinical practice. Next, we will detail each of the databases recommended by the standard, presenting specific characteristics of each of them.

7.5.1 MIT–BIH

Developed by a partnership between Beth Israel Hospital and the Massachusetts Institute of Technology (MIT), the MIT–BIH Database was the first standardized database for the evaluation of arrhythmia detectors to be available. The MIT–BIH is available since 1980 and has been distributed to more than 500 sites worldwide, being referenced in most cardiology-related publications. Initially it was distributed in digital tapes, but from 1989, a CD-ROM version of the database was produced and made available.

This database contains 48 ECG signal records of approximately 30 minutes each, belonging to 47 different patients, sampled at 360 Hz (see Fig. 7.3). It is unique in that it covers the five types of classes of arrhythmias proposed by the AAMI, as can be seen in Table 7.2. About 60% of the records were obtained from inpatients. Signals were extracted from 25 men aged 32–89 years and 22 women aged 23–89 years. Twenty-three of the records were randomly selected from an extensive library of 24-hour Holter tapes, and the remaining twenty-five records were specif-

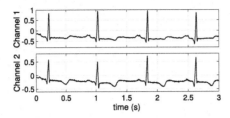

FIGURE 7.3

Record extracted from MIT–BIH.

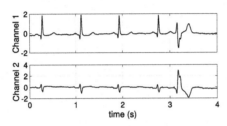

FIGURE 7.4

Record extracted from AHA.

ically selected to include a variety of important arrhythmias that would not be well represented by random sampling.

Each beat of the MIT–BIH Database was carefully analyzed by at least two cardiology specialists independently. After consensus regarding the class in which the beat belongs, annotations as per AAMI recommendations were performed. These notes (or labels) are used for the supervised learning of methods for automatic classification of arrhythmias, and also for evaluation of these. The recordings were digitized at 360 samples per second per channel, with 11-bit resolution, over a 10 mV range.

7.5.2 AHA

In 1982, almost simultaneously with the distribution of MIT–BIH, the researcher Ripley, G.C. Oliver and colleagues from the University of Washington, and other collaborating institutions, developed the American Heart Association's Database for Arrhythmia Detector Evaluation. However, the AHA database was designed specifically for the evaluation of ventricular arrhythmia detectors.

The AHA Database was developed in two parts, with eighty publicly released records and eighty kept "secret" for possible use by independent evaluation laboratories. Each database record contains two ECG leads, sampled at 250 Hz, and 12-bit resolution. The records lasted 3 hours and only the final 30 minutes were recorded, corresponding to beats; ventricular ectopic rhythms are noted (see Fig. 7.4).

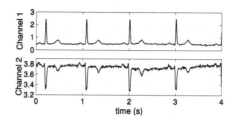

FIGURE 7.5

Record extracted from ESC ST–T.

These 80 records (designated as developmental set or series 1) are divided into eight classes of ten records each, according to the highest level of ventricular ectopy present:

- no ventricular ectopy (records 1001 through 1010);
- isolated unifocal PVCs (records 2001 through 2010);
- isolated multifocal PVCs (records 3001 through 3010);
- ventricular bi- and trigeminy (records 4001 through 4010);
- R-on-T PVCs (records 5001 through 5010);
- ventricular couplets (records 6001 through 6010);
- ventricular tachycardia (records 7001 through 7010);
- ventricular flutter/fibrillation (records 8001 through 8010).

7.5.3 ESC ST–T

The European Society of Cardiology ST–T Database is a database consisting of 90 records, each two hours long and contains two signals (that is, two leads). The records were sampled at 250 Hz and have a resolution of 12 bits (see Fig. 7.5). The database was originally constructed to allow analysis of the ST segment and T-wave, thus the ST–T ESC contains documented notes of individual beats, rhythm changes, noise episodes, and ST–T wave alterations. In cooperation with the MIT Center for Biomedical Engineering (developers of the MIT–BIH arrhythmia database), the annotation scheme was reviewed by two or more cardiologists to be consistent with the MIT–BIH and American Heart Association formats.

The records were acquired from 78 individuals, 70 men aged between 30 and 84 years, and 8 women who were between 55 and 71 years old. All of these subjects were suffering from some specific heart disease. The records include 372 ST and 423 T changes. Initially the complete development of the database was coordinated by the Institute of Clinical Physiology of the National Research Council (CNR), in Pisa, and the Thoraxcenter of Erasmus University in Rotterdam. However, in 1989 the European Society of Cardiology sponsored the remainder of the project, which was made available on CD-ROM in 1991.

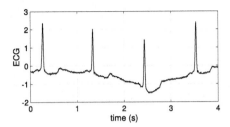

FIGURE 7.6

Record extracted from CU.

7.5.4 **CU**

The Creighton University Sustained Ventricular Arrhythmia Database consists of 35 records, each 8 minutes long, which were captured from individuals who had episodes of sustained ventricular tachycardia, ventricular flutter, and ventricular fibrillation. Each record contains 127,232 samples, were passed through an active second-order Bessel low-pass filter, with a cutoff of 70 Hz, and were digitized at 250 Hz with 12-bit resolution (see Fig. 7.6). The CU database provides reference annotation files to assist users in locating interest events in recordings.

In episodes of heart failure, a series of ventricular tachycardias almost always precedes fibrillation. The onset of fibrillation is extremely difficult to identify in many cases. Any detector must identify and respond to the series of tachycardia preceding fibrillation, since medical intervention is required as soon as possible (Nolle et al., 1986). Thus any detector that responds to premonitory tachycardia may exhibit a negative "alarm time" compared to the onset of fibrillation, as recorded in the reference annotation files. For this reason, the database is known as a tachyarrhythmia database, rather than a fibrillation database (Goldberger et al., 2000).

7.5.5 **NST**

Considering that noise is a major source of error in arrhythmia classifiers, the Noise Stress Test Database was developed with the objective of helping to observe the effects of noise in these systems, so that it could help to optimize the decisions of the algorithm based on signal quality. This database includes 12 half-hour ECG recordings and 3 typical half-hour noise recordings on ECG records found in ambulatory care. Examples of such interference are the electrode motion artefact, baseline wander, and muscle artefact (EMG).

Noise was artificially inserted into ECG signals available in the NST database, using physically active volunteers and standard ECG recorders, electrodes, and leads. The electrodes were inserted into the limbs, in positions where the ECGs of the individuals were not clearly detectable. The NST database was built based on two clean records of the MIT–BIH, and the noise was added to the database from the first 5 minutes of each record, during two-minute segments, alternating with clean two-minute segments. (See Fig. 7.7.)

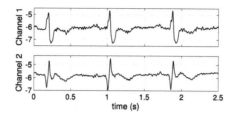

FIGURE 7.7

Record extracted from NST.

7.6 CONCLUSIONS

Many researches are being conducted focused on arrhythmia classification with ECG signals, and several authors have published promising results. However, there is much room for improvement, and consequently further research on the subject must be made. Among the algorithms of machine learning most used by the researchers, we highlighted the neural networks, support vector machines, and linear discriminants. However, the application of neural networks has stood out and presented better results, mainly in terms of accuracy. Unfortunately, the vast majority of studies in the literature still has not adopted a protocol that allows a fair comparison, as indicated in de Chazal et al. (2004) or standardizations of the AAMI standard. Recent techniques, such as deep learning have been applied for the detection of arrhythmias, showing satisfactory results. Studies investigating the application of real-time arrhythmia classification process are also being held. These research fields are promising and we encourage more work in this context.

The number of databases available for investigating and advancement of research that focus on fully automatic classification of heartbeats using ECG signals is still considered limited, and an obstacle to achieving further advances. Among the databases available for research with ECG signals, MIT–BIH is the most comprehensive, and used by researchers presenting all classes of signals cited by AMMI.

REFERENCES

Acharya, U.R., Oh, S.L., Hagiwara, Y., Tan, J.H., Adam, M., Gertych, A., San Tan, R., 2017. A deep convolutional neural network model to classify heartbeats. Computers in Biology and Medicine 89, 389–396.

Agrawal, A., Gawali, D.H., 2017. Comparative study of ECG feature extraction methods. In: 2017 2nd IEEE International Conference on Recent Trends in Electronics, Information Communication Technology (RTEICT), pp. 2021–2025.

Al Rahhal, M.M., Bazi, Y., AlHichri, H., Alajlan, N., Melgani, F., Yager, R.R., 2016. Deep learning approach for active classification of electrocardiogram signals. Information Sciences 345, 340–354.

Asl, B.M., Setarehdan, S.K., Mohebbi, M., 2008. Support vector machine-based arrhythmia classification using reduced features of heart rate variability signal. Artificial Intelligence in Medicine 44 (1), 51–64. https://doi.org/10.1016/j.artmed.2008.04.007. http://www.sciencedirect.com/science/article/pii/S0933365708000559.

Association for the Advancement of Medical Instrumentation, 1998. Testing and reporting performance results of cardiac rhythm and ST segment measurement algorithms. ANSI/AAMI EC38 1998.

Banupriya, C., Karpagavalli, S., 2014. Electrocardiogram beat classification using support vector machine and extreme learning machine. In: ICT and Critical Infrastructure: Proceedings of the 48th Annual Convention of Computer Society of India, vol. I. Springer, pp. 187–193.

Benjamin, E.J., Blaha, M.J., Chiuve, S.E., Cushman, M., Das, S.R., Deo, R., Floyd, J., Fornage, M., Gillespie, C., Isasi, C., et al., 2017. Heart disease and stroke statistics-2017 update: a report from the American Heart Association. Circulation 135 (10), e146–e603.

Bishop, C., 2006. Pattern Recognition and Machine Learning (Information Science and Statistics), pp. 138–147.

Bortolan, G., Christov, I., Pedrycz, W., 2007. Hyperbox classifiers for ECG beat analysis. In: Computers in Cardiology. 2007. IEEE, pp. 145–148.

Chawla, M.P.S., 2008. A comparative analysis of principal component and independent component techniques for electrocardiograms. Neural Computing & Applications 18, 539–556.

Chen, G., Wang, X., Wan, W., 2015. An ECG r-wave detection algorithm based on adaptive threshold. In: 2015 International Conference on Smart and Sustainable City and Big Data (ICSSC), pp. 145–149.

de Chazal, P., O'Dwyer, M., Reilly, R.B., 2004. Automatic classification of heartbeats using ECG morphology and heartbeat interval features. IEEE Transactions on Biomedical Engineering 51 (7), 1196–1206. https://doi.org/10.1109/TBME.2004.827359.

Doquire, G., De Lannoy, G., François, D., Verleysen, M., 2011. Feature selection for interpatient supervised heart beat classification. Computational Intelligence and Neuroscience 2011. https://doi.org/10.1155/2011/643816.

Escalona-Morán, M.A., Soriano, M.C., Fischer, I., Mirasso, C.R., 2015. Electrocardiogram classification using reservoir computing with logistic regression. IEEE Journal of Biomedical and Health Informatics 19 (3), 892–898.

Exarchos, T.P., Tsipouras, M.G., Exarchos, C.P., Papaloukas, C., Fotiadis, D.I., Michalis, L.K., 2007. A methodology for the automated creation of fuzzy expert systems for ischaemic and arrhythmic beat classification based on a set of rules obtained by a decision tree. Artificial Intelligence in Medicine 40 (3), 187–200. https://doi.org/10.1016/j.artmed.2007.04.001. http://www.sciencedirect.com/science/article/pii/S0933365707000504.

Gari, D.C., Francisco, A., Patrick, E., 2006. Advanced Methods and Tools for ECG Data Analysis. Artech House, Inc.

Gath, I., Inbar, G.F., 2013. Advances in Processing and Pattern Analysis of Biological Signals. Springer Science & Business Media.

Goldberger, A.L., Amaral, L.A., Glass, L., Hausdorff, J.M., Ivanov, P.C., Mark, R.G., Mietus, J.E., Moody, G.B., Peng, C.K., Stanley, H.E., 2000. PhysioBank, PhysioToolkit, and PhysioNet. Circulation 101 (23), e215–e220.

Gutiérrez-Gnecchi, J.A., Morfin-Magaña, R., Lorias-Espinoza, D., del Carmen Tellez-Anguiano, A., Reyes-Archundia, E., Méndez-Patiño, A., Castañeda-Miranda, R., 2017. DSP-based arrhythmia classification using wavelet transform and probabilistic neural network. Biomedical Signal Processing and Control 32, 44–56.

Halder, B., Mitra, S., Mitra, M., 2016. Detection and identification of ECG waves by histogram approach. In: 2016 2nd International Conference on Control, Instrumentation, Energy Communication (CIEC), pp. 168–172.

Hall, J.E., 2017. Tratado De Fisiologia Médica. Elsevier Brasil.

Jambukia, S.H., Dabhi, V.K., Prajapati, H.B., 2015. Classification of ECG signals using machine learning techniques: a survey. In: 2015 International Conference on Advances in Computer Engineering and Applications (ICACEA). IEEE, pp. 714–721.

Kanaan, L., Merheb, D., Kallas, M., Francis, C., Amoud, H., Honeine, P., 2011. PCA and KPCA of ECG signals with binary SVM classification. In: 2011 IEEE Workshop on Signal Processing Systems (SiPS), pp. 344–348.

Karpagachelvi, S., Arthanari, M., Sivakumar, M., 2010. ECG feature extraction techniques – a survey approach. CoRR. arXiv:1005.0957. http://arxiv.org/abs/1005.0957, 2010.

Khalaf, A.F., Owis, M.I., Yassine, I.A., 2015. A novel technique for cardiac arrhythmia classification using spectral correlation and support vector machines. Expert Systems with Applications 42 (21), 8361–8368.

Khazaee, A., Ebrahimzadeh, A., 2010. Classification of electrocardiogram signals with support vector machines and genetic algorithms using power spectral features. Biomedical Signal Processing and Control 5 (4), 252–263.

Khorrami, H., Moavenian, M., 2010. A comparative study of DWT, CWT and DCT transformations in ECG arrhythmias classification. Expert Systems with Applications 37 (8), 5751–5757.

Korürek, M., Doğan, B., 2010. ECG beat classification using particle swarm optimization and radial basis function neural network. Expert Systems with Applications 37 (12), 7563–7569. https://doi.org/10.1016/j.eswa.2010.04.087.

Korürek, M., Nizam, A., 2008. A new arrhythmia clustering technique based on ant colony optimization. Journal of Biomedical Informatics 41 (6), 874–881.

Krose, B., van der Smagt, Patrick, 1996. An Introduction to Neural Networks, pp. 16–18.

Kumar, R.G., Kumaraswamy, Y.S., 2013. Investigation and classification of ECG beat using input output additional weighted feed forward neural network. In: 2013 International Conference on Signal Processing, Image Processing & Pattern Recognition, pp. 200–205.

Lagerholm, M., Peterson, C., Braccini, G., Edenbrandt, L., Sornmo, L., 2000. Clustering ECG complexes using Hermite functions and self-organizing maps. IEEE Transactions on Biomedical Engineering 47 (7), 838–848.

Lanata, A., Valenza, G., Mancuso, C., Scilingo, E.P., 2011. Robust multiple cardiac arrhythmia detection through bispectrum analysis. Expert Systems with Applications 38 (6), 6798–6804.

Lim, C.L.P., Woo, W.L., Dlay, S.S., 2015. Enhanced wavelet transformation for feature extraction in highly variated ECG signal. In: 2nd IET International Conference on Intelligent Signal Processing 2015 (ISP), pp. 1–6.

Lin, C.C., Yang, C.-M., 2014. Heartbeat classification using normalized RR intervals and wavelet features. In: 2014 International Symposium on Computer, Consumer and Control, pp. 650–653.

Lin, S.W., Ying, K.C., Chen, S.C., Lee, Z.J., 2008. Particle swarm optimization for parameter determination and feature selection of support vector machines. Expert Systems with Applications 35 (4), 1817–1824. https://doi.org/10.1016/j.eswa.2007.08.088.

Llamedo, M., Martinez, J.P., 2011. Heartbeat classification using feature selection driven by database generalization criteria. IEEE Transactions on Biomedical Engineering 58 (3), 616–625. https://doi.org/10.1109/TBME.2010.2068048.

Luz, E.J.D.S., Nunes, T.M., De Albuquerque, V.H.C., Papa, J.P., Menotti, D., 2013. ECG arrhythmia classification based on optimum-path forest. Expert Systems with Applications 40 (9), 3561–3573. https://doi.org/10.1016/j.eswa.2012.12.063.

Luz, E.J.d.S., Schwartz, W.R., Cámara-Chávez, G., Menotti, D., 2016. ECG-based heartbeat classification for arrhythmia detection: a survey. Computer Methods and Programs in Biomedicine 127, 144–164.

Lynn, P., 1971. Recursive digital filters for biological signals. Medical & Biological Engineering 9 (1), 37–43.

Mar, T., Zaunseder, S., Martínez, J.P., Llamedo, M., Poll, R., 2011. Optimization of ECG classification by means of feature selection. IEEE Transactions on Biomedical Engineering 58 (8), 2168–2177.

Martis, R.J., Acharya, U.R., Adeli, H., 2014. Current methods in electrocardiogram characterization. Computers in Biology and Medicine 48, 133–149.

McCulloch, W.S., Pitts, W., 1943. A logical calculus of the ideas immanent in nervous activity. The Bulletin of Mathematical Biophysics 5 (4), 115–133.

Mishra, A.K., Raghav, S., 2010. Local fractal dimension-based ECG arrhythmia classification. Biomedical Signal Processing and Control 5 (2), 114–123.

Mukherjee, A., Ghosh, K.K., 2012. An efficient wavelet analysis for ECG signal processing. In: 2012 International Conference on Informatics, Electronics Vision (ICIEV), pp. 411–415.

Nasiri, J.A., Naghibzadeh, M., Yazdi, H.S., Naghibzadeh, B., 2009. ECG arrhythmia classification with support vector machines and genetic algorithm. In: 2009 Third UKSim European Symposium on Computer Modeling and Simulation. EMS'09. IEEE, pp. 187–192.

Nolle, F., Badura, F., Catlett, J., Bowser, R., Sketch, M., 1986. CREI-GARD, a new concept in computerized arrhythmia monitoring systems. Computers in Cardiology 13, 515–518.

Oh, I.S., Lee, J.S., Moon, B.R., 2004. Hybrid genetic algorithms for feature selection. IEEE Transactions on Pattern Analysis and Machine Intelligence 26 (11), 1424–1437. https://doi.org/10.1109/TPAMI.2004.105.

Osowski, S., Markiewicz, T., Hoai, L.T., 2008. Recognition and classification system of arrhythmia using ensemble of neural networks. Measurement 41 (6), 610–617.

Özbay, Y., Tezel, G., 2010. A new method for classification of ECG arrhythmias using neural network with adaptive activation function. Digital Signal Processing 20 (4), 1040–1049. https://doi.org/10.1016/j.dsp.2009.10.016.

Özbay, Y., Ceylan, R., Karlik, B., 2006. A fuzzy clustering neural network architecture for classification of ECG arrhythmias. Computers in Biology and Medicine 36 (4), 376–388.

Pan, J., Tompkins, W.J., 1985. A real-time QRS detection algorithm. IEEE Transactions on Biomedical Engineering BME-32 (3), 230–236. https://doi.org/10.1109/TBME.1985.325532.

Park, K., Cho, B., Lee, D., Song, S., Lee, J., Chee, Y., Kim, I., Kim, S., 2008. Hierarchical support vector machine based heartbeat classification using higher order statistics and Hermite basis function. In: Computers in Cardiology. 2008. IEEE, pp. 229–232.

Phegade, M., Mukherji, P., 2013. ICA based ECG signal denoising. In: 2013 International Conference on Advances in Computing, Communications and Informatics (ICACCI), pp. 1675–1680.

Proakis, J.G., 2001. Digital Signal Processing: Principles Algorithms and Applications. Pearson Education, India.

Qin, Q., Li, J., Zhang, L., Yue, Y., Liu, C., 2017. Combining low-dimensional wavelet features and support vector machine for arrhythmia beat classification. Scientific Reports 7 (1), 6067.

Rodríguez, R., Mexicano, A., Bila, J., Cervantes, S., Ponce, R., 2015. Feature extraction of electrocardiogram signals by applying adaptive threshold and principal component analysis. Journal of Applied Research and Technology 13, 261–269. http://www.scielo.org.mx/scielo.php?script=sci_arttext&pid=S1665-64232015000200012&nrm=iso.

Sarma, P., Nirmala, S.R., Sarma, K.K., 2013. Classification of ECG using some novel features. In: 2013 1st International Conference on Emerging Trends and Applications in Computer Science, pp. 187–191.

Schölkopf, B., Smola, A.J., 2002. Learning with Kernels: Support Vector Machines, Regularization, Optimization, and Beyond. MIT Press.

Shynk, J.J., 2012. Probability, Random Variables, and Random Processes: Theory and Signal Processing Applications. John Wiley & Sons.

da Silva, M.J., 2015. Characterization of QRS complex in ECG signals applying wavelet transform. In: 2015 International Conference on Mechatronics, Electronics and Automotive Engineering (ICMEAE), pp. 86–89.

Valluraiah, P., Biswal, B., 2015. ECG signal analysis using Hilbert transform. In: 2015 IEEE Power, Communication and Information Technology Conference (PCITC), pp. 465–469.

Vaneghi, F.M., Oladazimi, M., Shiman, F., Kordi, A., Safari, M.J., Ibrahim, F., 2012. A comparative approach to ECG feature extraction methods. In: 2012 Third International Conference on Intelligent Systems Modelling and Simulation, pp. 252–256.

Vaseghi, S.V., 2008. Advanced Digital Signal Processing and Noise Reduction. John Wiley & Sons.

Vincent, A.E., Sreekumar, K., 2017. A survey on approaches for ECG signal analysis with focus to feature extraction and classification. In: 2017 International Conference on Inventive Communication and Computational Technologies (ICICCT), pp. 140–144.

Wang, X., Yang, J., Teng, X., Xia, W., Jensen, R., 2007. Feature selection based on rough sets and particle swarm optimization. Pattern Recognition Letters 28 (4), 459–471. https://doi.org/10.1016/j.patrec.2006.09.003.

Ye, C., Coimbra, M.T., Kumar, B.V.K.V., 2010. Arrhythmia detection and classification using morphological and dynamic features of ECG signals. In: 2010 Annual International Conference of the IEEE Engineering in Medicine and Biology, pp. 1918–1921.

Yeh, Y.C., Wang, W.J., Chiou, C.W., 2009. Cardiac arrhythmia diagnosis method using linear discriminant analysis on ECG signals. Measurement 42 (5), 778–789.

Yu, S.N., Chen, Y.H., 2007. Electrocardiogram beat classification based on wavelet transformation and probabilistic neural network. Pattern Recognition Letters 28 (10), 1142–1150.

Zhang, Z., Dong, J., Luo, X., Choi, K.S., Wu, X., 2014. Heartbeat classification using disease-specific feature selection. Computers in Biology and Medicine 46, 79–89. https://doi.org/10.1016/j.compbiomed.2013.11.019.

Index

Printed in the United States
By Bookmasters